Mastering Python Scientific Computing

A complete guide for Python programmers to master
scientific computing using Python APIs and tools

Hemant Kumar Mehta

[PACKT] open source*

community experience distilled

PUBLISHING

BIRMINGHAM - MUMBAI

Mastering Python Scientific Computing

First published: September 2015

Production reference: 1180915

Published by Packt Publishing Ltd.
Livery Place
35 Livery Street
Birmingham B3 2PB, UK.

ISBN 978-1-78328-882-3

www.packtpub.com

Credits

Author
Hemant Kumar Mehta

Reviewers
Austen Groener
Sachin R. Joglekar

Commissioning Editor
Kartikey Pandey

Acquisition Editor
Kevin Colaco

Content Development Editor
Arshiya Umer

Technical Editor
Mohita Vyas

Copy Editor
Vikrant Phadke

Project Coordinator
Sanjeet Rao

Proofreader
Safis Editing

Indexer
Tejal Soni

Graphics
Jason Monteiro

Production Coordinator
Aparna Bhagat

Cover Work
Aparna Bhagat

About the Author

Hemant Kumar Mehta is a distributed and scientific computing enthusiast. He has more than 13 years of experience of teaching, research, and software development. He received his BSc (in computer science) Hons., master of computer applications degree, and PhD in computer science from Devi Ahilya University, Indore, India in 1998, 2001, and 2011, respectively. He has experience of working in diverse international environments as a software developer in MNCs. He is a post-doctorate fellow at an international university of high reputation.

Hemant has published more than 20 highly cited research papers in reputed national and international conferences and journals sponsored by ACM, IEEE, and Springer. He is the author of *Getting Started with Oracle Public Cloud, Packt Publishing*. He is also the coauthor of a book named *Internet and Web Technology*, published by Kaushal Prakashan Mandir, Indore.

He earned his PhD in the field of cloud computing and big data. Hemant is a member of ACM (Special Interest Group on High-performance Computing Education: SIGHPC-Edu), senior member of IEEE (the computer society, STC on cloud computing, and the big data technical committee), and a senior member of IACSIT, IAENG, and MIR Labs.

I am extremely thankful to my PhD supervisors, namely Professor Priyesh Kanungo and the late Professor Manohar Chandwani from Devi Ahilya University. Their words work as continuous guiding lights in my career and life.

I express heartfelt thanks to my dear student and friend, Pawan Pawar, for helping me develop some programs for this book.

I am also thankful to the entire Packt Publishing team and the reviewers for their tremendous support in maintaining the highest quality of work in this book.

Most of all, I thank my family. I am infinitely grateful to my parents. I thank my wife, Priya, and darling sons, Luv and Darsh, for whom this acknowledgement cannot be covered in words.

About the Reviewers

Austen Groener was raised in Southfield, Massachusetts, USA. He completed his BA in physics from Hartwick College and went on to pursue his MS and PhD in physics from Drexel University in Philadelphia, Pennsylvania, USA. He is a reputed astrophysicist, with research interests surrounding the detailed distribution of dark matter within the largest objects in the universe — galaxy clusters. When he is not studying the cosmos, he enjoys spending his free time developing software tools for other astronomers to use. Austen has a newfound interest in web development.

> I would like to thank my family and friends for their unwavering support. To my wife, Brittany: you are the love of my life, my best friend, and my inspiration.

Sachin R. Joglekar is a computer science graduate from BITS-Pilani (Goa campus) in India. His areas of interest primarily include machine learning and intelligent systems. He graduated in December 2014. Since then, he has been working as the cofounder of a start-up based in Mumbai. His work involves the design and development of server infrastructure and backend analytics for sensor networks. Sachin has also worked as an open source developer for SymPy, a symbolic computing library written in pure Python. His work at Google Summer of Code 2014 involved developing the vector module for SymPy.

www.PacktPub.com

Support files, eBooks, discount offers, and more

For support files and downloads related to your book, please visit www.PacktPub.com.

Did you know that Packt offers eBook versions of every book published, with PDF and ePub files available? You can upgrade to the eBook version at www.PacktPub.com and as a print book customer, you are entitled to a discount on the eBook copy. Get in touch with us at service@packtpub.com for more details.

At www.PacktPub.com, you can also read a collection of free technical articles, sign up for a range of free newsletters and receive exclusive discounts and offers on Packt books and eBooks.

https://www2.packtpub.com/books/subscription/packtlib

Do you need instant solutions to your IT questions? PacktLib is Packt's online digital book library. Here, you can search, access, and read Packt's entire library of books.

Why subscribe?

- Fully searchable across every book published by Packt
- Copy and paste, print, and bookmark content
- On demand and accessible via a web browser

Free access for Packt account holders

If you have an account with Packt at www.PacktPub.com, you can use this to access PacktLib today and view 9 entirely free books. Simply use your login credentials for immediate access.

To my parents and my gurus, Late Prof. Manohar Chandwani and Prof. Priyesh Kanungo

Table of Contents

Preface

"I am absolutely convinced that in a few decades, historians of science will describe the period we are in right now as one of deep and significant transformations to the very structure of science. And in that process, the rise of free openly available tools plays a central role."

Fernando Perez, creator of IPython

This book covers the Python APIs and toolkits used to perform scientific computing. It is highly recommended for readers who perform computerized engineering or scientific computations. Scientific computing is an interdisciplinary branch that requires a background in computer science, mathematics, general science (at least any one branch out of physics, chemistry, environmental science, biology, and others), and engineering. Python consists of a large number of packages, APIs, and toolkits for supporting the functionalities required by these diverse scientific and engineering domains.

A large community of users, lots of help and documentation, a large collection of scientific libraries and environments, great performance, and good support make Python a great choice for scientific computing.

What this book covers

Chapter 1, The Landscape of Scientific Computing – and Why Python?, introduces the basic concepts of scientific computing. It also discusses the background of Python, its guiding principle, and why using Python for scientific computing is efficient.

Chapter 2, A Deeper Dive into Scientific Workflows and the Ingredients of Scientific Computing Recipes, discusses the various concepts of mathematical and numerical analysis that are generally required to solve scientific problems. It also covers a brief introduction to the packages, toolkits, and APIs meant for performing scientific computing in the Python language.

Chapter 3, Efficiently Fabricating and Managing Scientific Data, discusses all the aspects about the underlying data of scientific applications, including the basic concepts, various operations, and the formats and software used to store data. It also presents standard datasets and techniques of preparing synthetic data.

Chapter 4, Scientific Computing APIs for Python, covers the basic concepts, features, and selected sample programs of various scientific computing APIs and toolkits, including NumPy, SciPy, and SymPy. A basic introduction to interactive computing, data analysis, and data visualization is also discussed in this chapter using IPython, matplotlib, and pandas.

Chapter 5, Performing Numerical Computing, discusses how to perform numerical computations using the NumPy and SciPy packages of Python. This chapter starts with the basics of numerical computation and covers a number of advanced concepts, such as optimization, interpolation, Fourier transformation, signal processing, linear algebra, statistics, spatial algorithms, image processing, file input/output, and others.

Chapter 6, Applying Python for Symbolic Computing, starts with the fundamentals of the Computerized Algebra System (CAS) and performing symbolic computations using SymPy. It covers a vast range of topics on CAS, from using simple expressions and basic arithmetic to advanced concepts of mathematics and physics.

Chapter 7, Data Analysis and Visualization, presents the concepts and applications of matplotlib and pandas for data analysis and visualization.

Chapter 8, Parallel and Large-scale Scientific Computing, discusses the concepts of high-performance scientific computing using IPython (which is done using MPI), the management of the Amazon EC2 cluster using StarCluster, multiprocessing, multithreading, Hadoop, and Spark.

Chapter 9, Revisiting Real-life Case Studies, illustrates several case studies of scientific computing applications, libraries, and tools developed using the Python language. Some cases studied from various engineering and science domains are presented in this chapter.

Chapter 10, *Best Practices for Scientific Computing*, discusses the best practices for scientific computing. It consists of the best practices for designing, coding, data management, application deployment, high-performance computing, security, data privacy, maintenance, and support. We also cover the best practices for general Python-based development.

What you need for this book

The example programs given in this book require a computer with Python 2.7.9 or a higher version, and several Python APIs/packages/toolkits. You will also require some Python libraries (namely NumPy, SciPy, SymPy, matplotlib, pandas, IPython), the IPython.parallel package, pyzmq, SSH for security (if necessary), and Hadoop.

Who this book is for

The book is intended for Python programmers willing to get hands-on exposure to scientific computing. The book expects that you have had exposure to various concepts of Python programming.

Conventions

In this book, you will find a number of text styles that distinguish between different kinds of information. Here are some examples of these styles and an explanation of their meaning.

Code words in text, database table names, folder names, filenames, file extensions, pathnames, dummy URLs, user input, and Twitter handles are shown as follows: "The functions of the random module are bound methods of a hidden instance of the random.Random class."

A block of code is set as follows:

```
import random
print random.random()
print random.uniform(1,9)
print random.randrange(20)
print random.randrange(0, 99, 3)
print random.choice('ABCDEFGHIJKLMNOPQRSTUVWXYZ') # Output 'P'
items = [1, 2, 3, 4, 5, 6, 7, 8, 9, 10]
random.shuffle(items)
print items
```

```
print random.sample([1, 2, 3, 4, 5, 6, 7, 8, 9, 10],  5)
weighted_choices = [('Three', 3), ('Two', 2), ('One', 1), ('Four', 4)]
population = [val for val, cnt in weighted_choices for i in
range(cnt)]
print random.choice(population)
```

Warnings or important notes appear in a box like this.

Tips and tricks appear like this.

Reader feedback

Feedback from our readers is always welcome. Let us know what you think about this book—what you liked or disliked. Reader feedback is important for us as it helps us develop titles that you will really get the most out of.

To send us general feedback, simply e-mail feedback@packtpub.com, and mention the book's title in the subject of your message.

If there is a topic that you have expertise in and you are interested in either writing or contributing to a book, see our author guide at www.packtpub.com/authors.

Customer support

Now that you are the proud owner of a Packt book, we have a number of things to help you to get the most from your purchase.

Downloading the example code

You can download the example code files from your account at http://www.packtpub.com for all the Packt Publishing books you have purchased. If you purchased this book elsewhere, you can visit http://www.packtpub.com/support and register to have the files e-mailed directly to you.

Downloading the color images of this book

We also provide you with a PDF file that has color images of the screenshots/ diagrams used in this book. The color images will help you better understand the changes in the output. You can download this file from `https://www.packtpub.com/sites/default/files/downloads/8823OS.pdf`.

Errata

Although we have taken every care to ensure the accuracy of our content, mistakes do happen. If you find a mistake in one of our books—maybe a mistake in the text or the code—we would be grateful if you could report this to us. By doing so, you can save other readers from frustration and help us improve subsequent versions of this book. If you find any errata, please report them by visiting `http://www.packtpub.com/submit-errata`, selecting your book, clicking on the **Errata Submission Form** link, and entering the details of your errata. Once your errata are verified, your submission will be accepted and the errata will be uploaded to our website or added to any list of existing errata under the Errata section of that title.

To view the previously submitted errata, go to `https://www.packtpub.com/books/content/support` and enter the name of the book in the search field. The required information will appear under the **Errata** section.

Piracy

Piracy of copyrighted material on the Internet is an ongoing problem across all media. At Packt, we take the protection of our copyright and licenses very seriously. If you come across any illegal copies of our works in any form on the Internet, please provide us with the location address or website name immediately so that we can pursue a remedy.

Please contact us at `copyright@packtpub.com` with a link to the suspected pirated material.

We appreciate your help in protecting our authors and our ability to bring you valuable content.

Questions

If you have a problem with any aspect of this book, you can contact us at `questions@packtpub.com`, and we will do our best to address the problem.

1
The Landscape of Scientific Computing – and Why Python?

Using computerized mathematical modeling and numerical analysis techniques to analyze and solve problems in the science and engineering domains is called **scientific computing**. Scientific problems include problems from various branches of science, such as earth science, space science, social science, life science, physical science, and formal science. These branches cover almost all the science domains that exist, from traditional science to modern engineering science, such as computer science. Engineering problems include problems from civil and electrical to (the latest) biomedical engineering.

In this chapter, we will cover the following topics:

- Fundamentals of scientific computing
- The flow of the scientific computation process
- Examples from scientific and engineering domains
- The strategy to solve complex problems
- Approximation, errors, and related terms
- Concepts of error analysis
- Computer arithmetic and floating-point numbers
- A background of Python
- Why choose Python for scientific computing?

Mathematical modeling refers to modeling activity that involves mathematical terms to represent the behavior of devices, objects, phenomena, and concepts. Generally, it helps in better understanding of the behavior or observations of a concept, a device, or objects. It may help explain the observation and possibly prediction of some future behavior, or results that are yet to be observed or measured. Numerical analysis is an area of computer science and mathematics that designs, analyzes, and finally implements algorithms to numerically solve problems of natural sciences (for example, physics, biology, and earth science), social sciences (for example, economics, psychology, sociology, and political science), engineering, medicine, and business. There is a package and workflow named **Python Dynamics** (**PyDy**) that is used to study multibody dynamics. It is a workflow and a software package developed on top of the SymPy mechanics package. PyDy extends SymPy and facilitates the simulation of multibody dynamics.

Definition of scientific computing

Scientific computing can also be called **computational science** or **scientific computation**. It is mainly the idea of development of mathematical models, use of quantitative analysis techniques, and use of computers for solving scientific problems.

> *"Scientific computing is the collection of tools, techniques and theories required to solve on a computer the mathematical models of problems in science and engineering."*

> *– Gene H. Golub and James M. Ortega*

In simple words, scientific computing can be described as an interdisciplinary field, as presented in the following diagram:

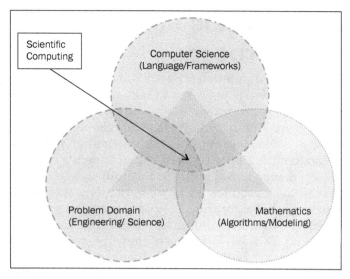

Scientific computing as an interdisciplinary field

Scientific computing requires knowledge of the subject of the underlying problem to be solved (generally, it will be a problem from a science or engineering domain), a mathematical modeling capability with a sound idea of various numerical analysis techniques, and finally its efficient and high-performance implementation using computing techniques. It also requires application of computers; various peripherals, including networking devices, storage units, processing units, and mathematical and numerical analysis software; programming languages; and any database along with a good knowledge of the problem domain. The use of computation and related technologies has enabled newer applications, and scientists can infer new knowledge from existing data and processes.

In terms of computer science, scientific computing can be considered a numerical simulation of a mathematical model and domain data/information. The objective behind a simulation depends on the domain of the application under simulation. The objective can be to understand the cause behind an event, reconstruct a specific situation, optimize the process, or predict the occurrence of an event. There are several situations where numerical simulation is the only choice, or the best choice. There are some phenomena or situations where performing experiments is almost impossible, for example, climate research, astrophysics, and weather forecasts. In some other situations, actual experiments are not preferable, for example, to check the stability or strength of some material or product. Some experiments are very costly in terms of time/economy, such as car crashes or life science experiments. In such scenarios, scientific computing helps users analyze and solve problems without spending much time or cost.

A simple flow of the scientific computation process

A simple flow diagram of computation for a scientific application is depicted in the next diagram. The first step is to design a mathematical model for the problem under consideration. After the formulation of the mathematical model, the next step is to develop its algorithm. This algorithm is then implemented using a suitable programming language and an appropriate implementation framework. Selecting the programming language is a crucial decision that depends on the performance and processing requirements of the application. Another close decision is to finalize the framework to be used for the implementation. After deciding on the language and framework, the algorithm is implemented and sample simulations are performed. The results obtained from simulations are then analyzed for performance and correctness. If the result or performance of the implementation is not as per expectations, its causes should be determined. Then we need to go back to either reformulate the mathematical model, or redesign the algorithm or its implementation and again select the language and the framework.

A mathematical model is expressed by a set of suitable equations that describe most problems to the right extent of details. The algorithm represents the solution process in individual steps, and these will be implemented using a suitable programming language or scripting.

After implementation, there is an important step to perform—the simulation run of the implemented code. This involves designing the experimentation infrastructure, preparing or arranging the data/situation for simulation, preparing the scenario to simulate, and much more.

After completing a simulation run, result collection and its presentation are desired for the next step to analyze the results so as to test the validity of the simulation. If the results are not as they are expected, then this may require going back to one of the previous steps of the process to correct and repeat them. This situation is represented in the following figure in the form of dashed lines going back to some previous steps. If everything goes ahead perfectly, then the analysis will be the last step of the workflow, which is represented by double lines in this diagram:

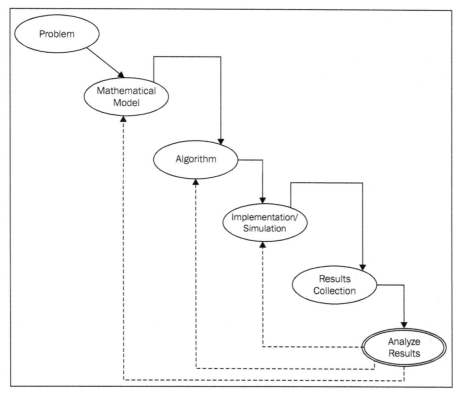

Various steps in a scientific computing workflow

The design and analysis of algorithms that solves any mathematical problem, specifically about science and engineering, is known as **numerical analysis**, and nowadays it is also called scientific computing. In scientific computing, the problems under consideration mainly deal with continuous values rather than discrete values. The latter are dealt with in other computer science problems. Generally saying, scientific computing solves problems that involve functions and equations with continuous variables, for example, time, distance, velocity, weight, height, size, temperature, density, pressure, stress, and much more.

Generally, problems of continuous mathematics have approximate solutions, as their exact solution is not always possible in a finite number of steps. Hence, these problems are solved using an iterative process that finally converges to an acceptable solution. The acceptable solution depends on the nature of the specific problem. Generally, the iterative process is not infinite, and after each iteration, the current solution gets closer to the desired solution for the purpose of simulation. Reviewing the accuracy of the solution and swift convergence to the solution form the gist of the scientific computing process.

There are well-established areas of science that use scientific computing to solve problems. They are as follows:

- Computational fluid dynamics
- Atmospheric science
- Seismology
- Structural analysis
- Chemistry
- Magnetohydrodynamics
- Reservoir modeling
- Global ocean/climate modeling
- Astronomy/astrophysics
- Cosmology
- Environmental studies
- Nuclear engineering

Recently, some emerging areas have also started harnessing the power of scientific computing. They include:

- Biology
- Economics
- Materials research
- Medical imaging
- Animal science

Examples from scientific/engineering domains

Let's take a look at some problems that may be solved using scientific computing. The first problem is to study the behavior of a collision of two black holes, which is very difficult to understand theoretically and practically. Theoretically, this process is extremely complex, and it is almost impossible to perform it in a laboratory and study it live. But this phenomenon can be simulated in a computing laboratory with a proper and efficient implementation of a mathematical formulation of Einstein's general theory of relativity. However, this requires very high computational power, which can be achieved using advanced distributed computing infrastructure.

The second problem is related to engineering and designing. Consider a problem related to automobile testing called **crash testing**. To reduce the cost of performing a risky actual crash for testing, engineers and designers prefer to perform a computerized simulated crash test. Finally, consider the problem of designing a large house or factory. It is possible to construct a dummy model of the proposed infrastructure. But that requires a reasonable amount of time and is expensive. However, this designing can done using an architectural design tool, and this will save a lot of time and cost. There can be similar examples from bioinformatics and medical science, such as protein structure folding and modeling of infectious diseases. Studying protein structure folding is a very time-consuming process, but it can be efficiently completed using large-scale supercomputers or distributed computing systems. Similarly, modeling an infectious disease will save efforts and cost in the analysis of the effects of various parameters on a vaccination program for that disease.

These three examples are selected as they represent three different classes of problems that can be solved using scientific computing. The first problem is almost impossible. The second problem is possible, but it is risky up to a certain extent and it may result in severe damage. The final problem can be solved without any simulation and it is possible to duplicate it in real-life situations. However, it is costlier and more time-consuming than its simulation.

A strategy for solving complex problems

A simple strategy to find a solution for a complex computational problem is to first identify the difficult areas in the solution. Now, one by one, start replacing these small difficult parts with their solutions that will lead to the same solution or to a solution within the problem-specific permissible limit. In other words, the best idea is to reduce a large, complex problem to a set of smaller problems. Each of them may be complex or simple. Now each of the complex subproblems may be replaced with a similar and simple problem, and in this way, we ultimately get a simpler problem to solve. The basic idea is to combine the divide-and-conquer technique with the change of smaller complex problems with similar simple problems.

We should take care of two important points when adopting this idea. The first is that we need to search for a similar problem or a problem that has a solution from the same class. The second is that just after the replacement of one problem with another, we need to determine whether the ultimate solution is preserved within the tolerance limit, if not completely preserved. Some examples may be as follows:

- Changing infinite-dimensional spaces in the problem to finite-dimensional spaces for simplicity

- Change infinite processes with finite processes, such as replacing integrals or infinite series with finite summations or a derivative of finite differences

- If feasible, then algebraic equations can be used to replace differential equations

- Try replacing nonlinear problems with linear problems as linear problems are very simple to solve

- If feasible, complicated functions can be changed to multiple simple functions to achieve simplicity

Approximation, errors, and associated concepts and terms

These scientific computational solutions generally produce approximate solutions. By approximate solution, we mean that instead of the exact desired solution, the obtained solution will be nearly similar to it. By nearly similar, we mean that it will be a sufficiently close solution to consider the practical or simulation successful, as they fulfill the purpose. This approximate, or similar, solution is caused by a number of sources. These sources can be divided into two categories: sources that arise before the computations begin, and those that occur during the computations.

The approximations that occur before the beginning of computations may be caused by one or more of the following:

- **Assumption or ignorance during modeling**: There might be an assumption during the modeling process, and similarly ignorance or omission of the impact of a concept or phenomenon during modeling, that may result in the approximation or tolerable inaccuracy.

- **Data derived from observations or experiments**: The inaccuracy may be in the data obtained from some devices that have low precision. During the computations, there are some constants, such as pi, whose values have to be approximated, and this is also an important cause of deviation from the correct result.

- **Prerequisite computations**: The data may have been obtained from the results of previous experiments, or simulations may have had minor, acceptable inaccuracies that finally led to further approximations. Such prior processing may be a prerequisite of the subsequent experiments.

Approximation during computations occurs because of one or more of the following sources:

- **Simplification of the problem**: As we have already suggested in this chapter, to solve large and complex problems, we should use a combination of "divide and conquer" and replacing a small, complex problem with a simpler one. This may result in approximations. Considering that we replaced an infinite series with a finite series will possibly cause approximations.

- **Truncation and rounding**: A number of situations ask for rounding and truncation of the intermediate results. Similarly, the internal representation of floating-point numbers in computers and their arithmetic also leads to minor inaccuracies.

The approximate value of the final result of a computation problem may be the outcome of any combination of the various sources discussed previously. The accuracy of the final output may be reduced or increased depending on the problem being solved and the approach used to solve it.

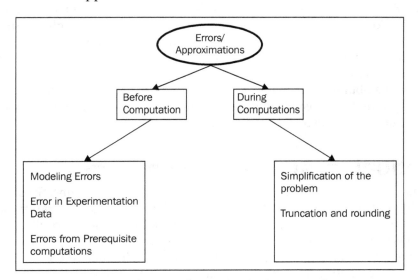

Taxonomy of errors and approximation in computations

Error analysis

Error analysis is a process used to observe the impact of such approximations on the accuracy of an algorithm or computational process. In the subsequent text, we are going to discuss the basic concepts associated with error analysis.

An observation may be made from the previous discussion on approximations that the errors can be considered as errors in the input data and they arose during the computations on this input data.

On a similar path, computation errors may again be divided into two categories: truncation errors and rounding errors. A truncation error is the result of reducing a complex problem to a simpler problem, for example, immature termination of iterations before the desired accuracy is achieved. A rounding error is the result of the precision used to represent numbers in the number system used for the computerized computation, and also the result of performing arithmetic on these numbers.

Ultimately, the amount of error that is significant or ignorable depends on the scale of the values. For example, an error of 10 in a final value of 15 is highly significant, while an error of 10 in a final value of 785 is not that significant. Moreover, the same error of 10 in obtaining the final value of 17,685 is ignorable. Generally, the impact of an error value is relative to the value of the result. If we know the magnitude of the final value to be obtained, then after looking at the value of the error, we can decide whether to ignore it or consider it as significant. If the error is significant, then we should start taking the corrective measures.

Conditioning, stability, and accuracy

Let's discuss some important properties of problems and algorithms. Sensitivity or conditioning is a property of a problem. The problem under consideration can be called sensitive or insensitive, or it may be called well-conditioned or ill-conditioned. A problem is said to be insensitive or well-conditioned if, for a given relative change in input, the data will have a proportional relative final impact on the result. On the other hand, if the relative impact of the final result is considerably larger than the relative change in input data, then the problem will be considered a sensitive or ill-conditioned problem.

Backward and forward error analysis

Assume that we have obtained the approximation y^* by f mapping the data x, for example, $y^*=f(x)$. Now, if the actual result is y, then the small quantity $y' =y^*-y$ is called a **forward error**, and its estimation is called forward error analysis. Generally, it is very difficult to obtain this estimate. An alternative approach to this is to consider y^* as the exact solution to the same problem with modified data, that is, $y^*=f(x')$. Now, the quantity $x^*=x'-x$ is called a backward error in y^*. Backward error analysis is the process of estimation of x^*.

Is it okay to ignore these errors?

The answer to this question depends on the domain and application where you are going to apply the scientific computations. For example, if it is the calculation of the time to launch a missile, an error of 0.1 seconds will result in severe damage. On the other hand, if it is the calculation of the arrival time of a train, an error of 40 seconds will not lead to a big problem. Similarly, a small change in a medicine dosage can have a disastrous effect on the patient. Generally, if a computation error in an application is not related to loss of human lives or doesn't involve big costs, then it can be ignored. Otherwise, we need to take proper efforts to resolve the issue.

Computer arithmetic and floating-point numbers

A type of approximation in scientific computing is introduced due to the representation of real numbers in computers. This approximation is further magnified by performing arithmetic operations on these real numbers. In this section, we will discuss this representation of real numbers, arithmetic operations on these numbers, and their possible impact on the results of the computation. These approximation errors not only arise in computerized computations, however; they may arise in non-computerized manual computation because of the rounding done to reduce the complexity. However, it is not the case that these approximations arise only in the case of computerized computations. They can also be observed in non-computerized, manual computations because of rounding done to reduce complexities in calculations.

Before advancing the discussion of the computerized representation of real numbers, let's first recall the well-known scientific notation used in mathematics. In scientific notation, to simplify the representation of a very large or very small number into a short form, we write nearly the same quantity multiplied by some powers of 10. Also, in scientific notation, numbers are represented in the form of "*a* multiplied by *10* to the power *b*" that is, a X 10b. For example, 0.000000987654 and 987,654 can represented as *9.87654 x 10^-7* and *9.87654 x 10^5* respectively. In this representation, the exponent is an integer quantity and the coefficient is a real number called **mantissa**.

The **Institute of Electrical and Electronics Engineers** (**IEEE**) has standardized the floating-point number representation in *IEEE 754*. Most modern machines use this standard as it addresses most of the problems found in various floating-point number representations. The latest version of this standard is published in 2008 and is known as *IEEE 754-2008*. The standard defines arithmetic formats, interchange formats, rounding rules, operations, and exception handling. It also includes recommendations for advanced exception handling, additional operations, and evaluation of expressions, and tells us how to achieve reproducible results.

The background of the Python programming language

Python is a general-purpose high-level programming language that supports most programming paradigms, including procedural, object-oriented, imperative, aspect-oriented, and functional programming. It also supports logical programming using an extension. It is an interpreted language that helps programmers compose a program in fewer lines than the code for the same concept in C++, Java, or other languages. Python supports dynamic typing and automatic memory management. It has a large and comprehensive standard library, and now it also has support for a number of custom libraries for many specific tasks. It is very easy to install packages using package managers such as `pip`, `easy_install`, `homebrew` (OS X), `apt-get` (Linux), and others.

Python is an open source language; its interpreters are available for most operating systems, including Windows, Linux, OS X, and others. There are a number of tools available to convert a Python program into an executable form for different operating systems, for example, Py2exe and PyInstaller. This executable form is standalone code that does not require a Python interpreter for execution.

The guiding principles of the Python language

Python's guiding principles by Guido van Rossum, who is also known as the **Benevolent Dictator For Life (BDFL)**, have been converted into some aphorism by Tim Peters and are available at `https://www.python.org/dev/peps/pep-0020/`. Let's discuss these with some explanations, as follows:

- **Beautiful is better than ugly**: The philosophy behind this is to write programs for human readers, with simple expression syntax and consistent syntax and behavior for all programs.

- **Explicit is better than implicit**: Most concepts are kept explicit, just like the explicit Boolean type. We have used an explicit literal value — true or false — for Boolean variables instead of depending on zero or nonzero integers. Still, it does support the integer-based Boolean concept. Nonzero values are treated as Boolean. Similarly, its `for` loop can operate data structures without managing the variable. The same loop can iterate through tuples and characters in a string.

- **Simple is better than complex**: Memory allocation and the garbage collector manage allocation or deallocation of memory to avoid complexity. Another simplicity is introduced in the simple print statement. This avoids the use of file descriptors for simple printing. Moreover, objects automatically get converted to a printable form in comma-separated values.

- **Complex is better than complicated**: Scientific computing concepts are complex, but this doesn't mean that the program will be complicated. Python programs are not complicated, even for very complex application. The "Pythonic" way is inherently simple, and the SciPy and NumPy packages are very good examples of this.

- **Flat is better than nested**: Python provides a wide variety of modules in its standard library. Namespaces in Python are kept in a flat structure, so there is no need to use very long names, such as `java.net.socket` instead of a simple socket in Python. Python's standard library follows the *batteries included* philosophy. This standard library provides tools suitable for many tasks. For example, modules for various network protocols are supported for the development of rich Internet applications. Similarly, modules for graphic user interface programming, database programming, regular expressions, high-precision arithmetic, unit testing, and much more are bundled in the standard library. Some of the modules in the library include networking (`socket`, `select`, `SocketServer`, `BaseHTTPServer`, `asyncore`, `asynchat`, `xmlrpclib`, and `SimpleXMLRPCServer`), Internet protocols (`urllib`, `httplib`, `ftplib`, `smtpd`, `smtplib`, `poplib`, `imaplib`, and `json`), database (`anydbm`, `pickle`, `shelve`, `sqlite3`, and `mongodb`), and parallel processing (`subprocess`, `threading`, `multiprocessing`, and `queue`).

- **Sparse is better than dense**: The Python standard library is kept shallow and the Python package index maintains an exhaustive list of third-party packages meant for supporting in-depth operations for a topic. We can use `pip` to install custom Python packages.

- **Readability counts**: The block structure of your program should be created using white spaces, and Python uses minimal punctuation in its syntax. As semicolons introduce blocks, no semicolons are needed at the end of the line. Semicolons are allowed but they are not required in every line of code. Similarly, in most situations, parentheses are not required for expressions. Python introduces inline documentation used to generate API documentation. Python's documentation is available at runtime and online.

- **Special cases aren't special enough to break the rules**: The philosophy behind this is that everything in Python is an object. All built-in types are implemented as objects. The data types that represent numbers have methods. Even functions are themselves objects with methods.

- **Although practicality beats purity**: Python supports multiple programming styles to give users the choice to select the style that is most suitable for their problem. It supports OOP, procedural, functional, and many more types of programming.

- **Errors should never pass silently**: It uses the concept of exception handling to avoid handling errors at low level APIs so that they may be handled at a higher level while writing the program that uses these APIs. It supports the concept of standard exceptions with specific meanings, and users are allowed to define exceptions for custom error handling. To support debugging of code, the concept of traceback is provided. In Python programs, by default, the error handling mechanism prints a complete traceback pointing to the error in `stderr`. The traceback includes the source filename, line number, and source code, if it is available.

- **Unless explicitly silenced**: To take care of some situations, there are options to let an error pass by silently. For these situations, we can use the `try` statement without `except`. There is also an option to convert an exception into a string.

- **In the face of ambiguity, refuse the temptation to guess**: Automatic type conversion is performed only when it is not surprising. For example, an operation between an integer operand with a float operand results in a float value.

- **There should be one—and preferably only one—obvious way to do it**: This is very obvious. It requires elimination of all redundancy. Hence, it is easier to learn and remember.

- **Although that way may not be obvious at first unless you're Dutch**: The way that we discussed in the previous point is applicable to the standard library. Of course, there will be redundancy in third-party modules. For example, we have support for multiple GUI APIs, such as as GTK, wxPython, and KDE. Similarly for web programming, we have Django, AppEngine, and Pyramid.

- **Now is better than never**: This statement is meant to motivate users to adopt Python as their favorite tool. There is a concept of ctypes meant to wrap existing C/C++ shared libraries for use in Python programs.

- **Although never is often better than *right* now**: With this philosophy, the **Python Enhancement Proposals** (**PEP**) processed a temporary moratorium (suspension) on all changes to the syntax, semantics, and built-in components for a specified period to promote the alternative development catch-up.

- **If the implementation is hard to explain, it's a bad idea** and **If the implementation is easy to explain, it may be a good idea**: In Python all the changes to the syntax, new library modules, and APIs will be processed through a highly rigorous process of review and approval.

Why Python for scientific computing?

To be frank, if we're talking about the Python language alone, then we need to think about some option. Fortunately, we have support for NumPy, SciPy, IPython, and matplotlib, and this makes Python the best choice. We are going to discuss these libraries in subsequent chapters. The following are the comprehensive features of Python and the associated library that make Python preferable to the other alternatives such as MATLAB, R, and other programming languages. Mostly, there is no single alternative that possesses all of these features.

Compact and readable code

Python code is generally compact and inherently more readable in comparison to its alternatives for scientific computing. As discussed in the Python guiding principles, this is the impact of the design philosophy of Python.

Holistic language design

Overall, the design of the Python language is highly convenient for scientific computing because Python supports multiple programming styles, including procedural, object-oriented, functional, and logic programming. The user has a wide range of choices and they can select the most suitable one for their problem. This is not the case with most of the available alternatives.

Free and open source

Python and the associated tools are freely available for use, and they are published as open source tools. This brings an added advantage of availability of their internal source code. On the other hand, most competing tools are costly proprietary products and their internal algorithms and concepts are not published for users.

Language interoperability

Python supports interoperability with most existing technologies. We can call or use functions, code, packages, and objects written in different languages, such as MATLAB, C, C++, R, Fortran, and others. There are a number of options available to support this interoperability, such as Ctypes, Cython, and SWIG.

Portable and extensible

Python supports most platforms. So, it is a portable programming language, and its program written for one platform will result in almost the same output on any other platform if Python toolkits are available for that platform. The design principles behind Python have made it a highly extensible language, and that's why we have a large number of high-class libraries available for a number of different tasks.

Hierarchical module system

Python supports a modular system to organize programs in the form of functions and classes in a namespace. The namespace system is very simple in order to keep learning and remembering the concepts easy. This also supports enhanced code reusability and maintenance.

Graphical user interface packages

The Python language offers a wide set of choices in graphics packages and tool sets. These toolkits and packages support graphic design, user interface designing, data visualization, and various other activities.

Data structures

Python supports an exhaustive range of data structures, which is the most important component in the design and implementation of a program to perform scientific computations. Support for a dictionary is the most highlightable feature of the data structure functionality of the Python language.

Python's testing framework

Python's unit testing framework, named PyUnit, supports complete unit testing functionality for integration with the `mypython` program. It supports various important unit testing concepts, including test fixture, test cases, test suites, and test runner.

Available libraries

Owing to the batteries-included philosophy of Python, it supports a wide range of standard packages in its bundled library. As it is an extensible language, a number of well-tested custom-specific purpose libraries are available for a wide range of users. Let's briefly discus a few libraries used for scientific computations.

NumPy/SciPy is a package that supports most mathematical and statistical operations required for any scientific computation. The SymPy library provides functionality for symbolic computations of basic symbolic arithmetic, algebra, calculus, discrete mathematics, quantum physics, and more. PyTables is a package used to efficiently process datasets that have a large amount of data in the form of a hierarchical database. IPython facilitates the interactive computing feature of Python. It is a command shell that supports interactive computing in multiple programming languages. matplotlib is a library that supports plotting functionality for Python/NumPy. It supports plotting of various types of graphs, such as line plot, histogram, scatter plot, and 3D plot. SQLAlchemy is an object-relational mapping library for Python programming. By using this library, we can use the database capability for scientific computations with great performance and ease. Finally, it is time to introduce a toolkit written on top of the packages we just discussed and a number of other open source libraries and toolkits. This toolkit is named SageMath. It is a piece of open source mathematical software.

The downsides of Python

After discussing a lot of upsides of Python over the alternatives, if we start searching for some downsides, we will notice something important: the **integrated development environment** (IDE) of Python is not the most powerful IDE compared to the alternatives. As Python toolkits are arranged in the form of discrete packages and toolkits, some of them have a command-line interface. So, in the comparison of this feature, Python is lagging behind some alternatives on specific platforms, for example, MATLAB on Windows. However, this doesn't mean that Python is not that convenient; it is equally comparable and supports ease of use.

Summary

In this chapter, we discussed the basic concepts of scientific computing and its definitions. Then we covered the flow of the scientific computing process. Next, we briefly discussed some examples from a few science and engineering domains. After the examples, we explained an effective strategy to solve complex problems. After that, we covered the concept of approximation, errors, and related terms.

We also discussed the background of the Python language and its guiding principles. Finally, we discussed why Python is the most suitable choice for scientific computing.

In the next chapter, we will discuss various mathematical/numerical analysis concepts involved in scientific computing. We will also cover various Python packages, toolkits, and APIs for scientific computing.

2
A Deeper Dive into Scientific Workflows and the Ingredients of Scientific Computing Recipes

Scientific workflow is a term used to describe a series of structured activities and computational steps required to solve a scientific computing problem. The computations involved in scientific computing are very intense, are highly complex, and also possess complicated dependencies. In the rest of the chapter, we will continue to use scientific computation problem words to represent a scientific workflow. Let's discuss the various mathematical and computing components required by most scientific computing problems.

In this chapter, we will cover the following topics:

- Mathematical components of scientific computations
- Scientific computing libraries of Python
- An introduction to NumPy
- An introduction to SciPy
- Data analysis using pandas
- **Interactive Python (IPython)** for interactive programming
- Symbolic computing using SymPy
- Data visualization using Matplotlib

Mathematical components of scientific computations

First, we will briefly discuss the various mathematical components that may occur in a scientific computation problem. We will also look for possible methods of solving problems. However, in this discussion, we will not look into the details of any method. In the later part, we will discuss the Python APIs relevant to these concepts.

A system of linear equations

The most common mathematical component that arises in most applications of scientific computing and applied mathematics is the system of linear algebraic equations. Generally, this system may occur due to approximations of nonlinear equations by linear equations or differential equations by algebraic equations.

A system of linear equations is a collection of simultaneous linear equations involving a set of variables, like this for example:

```
2 x1  +  1 x2  + 1 x3  =  1
1 x1  -  2 x2  - 1 x3  =  2
1 x1  +  1 x2  + 2 x3  =  2
```

This is a system of three linear equations with three unknown variables: x1, x2, and x3. A solution for this system is the assignment of numbers to these unknown variables such that the values satisfy all three equations simultaneously. The solution for these equations is shown as follows:

```
x1  =  (1/2)
x2  =  (-3/2)
x3  =  (3/2)
```

This solution satisfies all three equations. This is why we call these linear equations a system of linear equations—the equations are supposed to be considered as a system rather than individual equations. Generally, iterative methods are methods that require repetition of steps to compute the solution. In programming, this repetition is performed using any of the available loops. On the other hand, a non-iterative method uses computational formulas to find the solution. There is a wide variety of methods of solving systems of linear equations. There are iterative and non-iterative methods. For example, the Gaussian LU-factorization method and elimination method are two examples of non-iterative methods. The Jacobi iteration method and the Gauss-Seidel iteration method are two popular iterative methods.

A system of nonlinear equations

A nonlinear system of equations is a set of simultaneous equations in which the unknown variables appear as polynomials of degree higher than 1. The system can be single-dimensional or multi-dimensional. Generally, a linear equation is of the following form. For a given function f, we need to find the value x for which this condition is true:

f(x) = 0

This value of x is called the **root** or zero of the equation.

There are two different cases when solving linear equations, as follows. In the first case, there is a single nonlinear equation with one variable:

f: RàR (scalar)

The solution for such equation is a scalar x for which `f(x)` = 0. The second case is a system of n nonlinear equations with n unknown variables:

f: Rn à Rn (vector)

The solution for such types of equations is a vector x for which all components of the function f are simultaneously zero for all `f(x)` = 0.

For example, a one-dimensional nonlinear equation is as follows:

3x + sin(x) -ex = 0

Its approximate solution up to two decimal digits is `0.36`. An example of a system of nonlinear equations in two dimensions is given here:

3-x2=y
x+1=y

The solution vectors for the preceding system are `[1, 2]` and `[-2, -1]`.

There are a number of methods of solving nonlinear equations and systems of nonlinear equations. For one-dimensional equations, the methods are listed as follows:

- Bisection method
- Newton's method
- Secant method
- Interpolation method
- Inverse interpolation method
- Inverse quadratic interpolation
- Linear fractional interpolation

Similarly, for a system of nonlinear equations, again we have a number of methods, as follows:

- Newton's method
- Secant updating method
- Damped Newton's method
- Broyden's method

As these methods are iterative methods, their rate of convergence is an important property to be observed. By convergence, we mean that these methods start with an approximate solution and proceed towards obtaining the exact solution. The speed of converging towards a solution is known as the rate of convergence. A faster converging method is better for obtaining the exact solution as it will take less time. For some faster methods, such as Newton's method, choosing the initial approximation is an important step. There is always a possibility that some methods may not converge to the solution if their initial approximation is not close enough to the solution. There are some hybrid methods as a trade-off between performance and guaranteed solution; the damped Newton's method is an example of such a method. The SciPy package has implementations of a number of methods for solving systems of nonlinear equations. You can refer to the `http://docs.scipy.org/doc/scipy-0.14.0/reference/generated/scipy.optimize.newton.html` to get more information about the Newton-Raphson method and its implementation.

Optimization

Optimization is the process of trying to obtain the best possible solution. Generally, it will be the solution that has the maximum or minimum values among all. For example, if we need to know the cost of any project an organization is working on, then the option that gives minimum cost will be the optimized option. Similarly, if the comparison is among various sales strategies, then the strategy that produces maximum profit will be the optimized option. SciPy has a package for optimization techniques. You can refer to `http://docs.scipy.org/doc/scipy/reference/optimize.html` for more details and the implementation of these methods. Optimization has applications in various science and engineering domains. Some of these are as follows:

- Mechanics and engineering
- Economics
- Operations research
- Control engineering
- Petroleum engineering
- Molecular modeling

Interpolation

In science and engineering domains, users have a number of data points obtained from sampling or some experimentation. These data points represent the values of a function for particular values of the independent variable. Now, it is often a requirement to estimate the value of this function for the remaining points within the range. This process of estimation is called interpolation. It may be achieved by curve fitting or regression analysis.

For example, consider the following values of an independent variable x and the corresponding values of the function f:

```
x     4    5    6    7    8    9    10
f(x)  48        75   108  147  192  243  300
```

Using the interpolation methods, we can estimate the value of this function for other values of the variable, such as x=7.5, that is, f(7.5) or f(5.25). Although the function given in the preceding example is very simple (f(x) = 3x2), it may be any function from real-life examples. For example, this function may be the temperature reading of a server room of an Internet data center of an e-commerce organization. These temperature readings may be taken at some different points in time. The time for a temperature reading may be a fixed time interval between two readings, or it may be completely random. In this example, the function is the temperature of the server room for discrete values of independent, variable time. We need to estimate, or interpolate, the temperature for the remaining time of the day.

Another example can be as follows: the function is the average number of hours in a day that the users under study invest/waste in using Facebook or WhatsApp based on age. Based on this data, we can estimate the number of hours of Facebook or WhatsApp usage by users of some age other than the age in the data points.

Extrapolation

Another closely related problem is extrapolation. As the name suggests, it is the extension of interpolation for an estimation beyond the range of values of the independent variable. For example, if we have been given the values of the number of hours of Facebook/WhatsApp usage for the values of ages of users from 12 years to 65 years, then the problem of estimating the number of hours spent by users who are less than 12 years old and more than 65 years old comes under the scope of extrapolation. This is because it is beyond the range of the given data points of independent variables.

We have a number of methods available for both interpolation and extrapolation. The following are the names of some methods of interpolation:

- Piecewise constant interpolation
- Linear interpolation
- Polynomial interpolation
- Spline interpolation
- Interpolation via Gaussian processes

The followings are some methods of extrapolation:

- Linear extrapolation
- Polynomial extrapolation
- Conic extrapolation
- French curve extrapolation

Numerical integration

Numerical integration is the process of approximating the values of an integral using any numerical techniques. The numerical computation of an integral is also called a quadrature. We need to approximate numerical integration as there are some functions that cannot be analytically integrated. Even if a formula exists, it may not be the most efficient way of calculating the integral. In some situations, we are supposed to integrate an unknown function for which only some samples of the function are known. Using numerical integration, we approximate the values of the definite integrals. This is also based on polynomial fitting through a specified set of points and then integrating the approximating function. In Python, the SciPy package has a module for integration. For details about the integration module and its implementation, refer to `http://docs.scipy.org/doc/scipy/reference/integrate.html`. There are a number of methods of solving numerical integration, as follows:

- Simpson's rule
- Trapezoidal rule
- Refined trapezoidal rule
- Gaussian quadrature rule
- Newton-Cotes quadrature rule
- Gauss-Legendre integration

Numerical differentiation

Numerical differentiation is the process of estimating the derivative of a function using the known values of the function. It is highly useful in several situations. Generally, there are situations wherein we are not well aware whether an underlying function exists or not and we only have a discrete dataset. In such situations, users are interested in studying changes in the data that are related to the derivatives. Sometimes, for the sake of performance and simplicity, we prefer to approximate the derivative instead of computing its exact value, as the exact formulas are available but they are very complicated to solve. Differentiation is frequently used to solve optimization problems. Machine learning techniques also depend on numerical differentiation most of the time.

The methods of numerical differentiation are as follows:

- Finite difference approximation
- Differential quadrature
- Finite difference coefficients
- Differentiation by interpolation

Differential equations

A differential equation is a mathematical equation that can relate a function to its derivative. The function represents a physical quantity, the derivative corresponds to the rate of change of this quantity, and the equation is the relationship between the two. For example, the motion of a freely falling object under the force of gravity is generally represented using a set of differential equations. Differential equations have applications in a wide range of fields, including pure and applied mathematics, physics, engineering, and other subjects. Mainly, these subjects are concerned with various types of differential equations.

Differential equations are mainly used to model every physical, technical, and biological process. In many situations, differential equations may not be directly solvable. Hence, the solutions should be approximated using numerical methods. Most fundamental laws of physics (for example, Newton's second law and Einstein's field equations) and chemistry, such as the rate law or rate equation, have been formulated as differential equations. Differential equations have been used to model the behavior of complex systems in biology (for example, biological population growth) and economics (for example, the simple exponential growth model).

Differential equations can be categorized into two types: **ordinary differential equations** (ODE) and **partial differential equations** (PDE). An ODE is an equation that contains a function of one independent variable and its derivatives. On the other hand, a PDE contains functions of multiple independent variables and their partial derivatives. The partial derivative of a function with multiple variables is the derivative of this function with respect to one of the variables. You may refer to `http://docs.scipy.org/doc/scipy-0.13.0/reference/generated/scipy.integrate.ode.html` for the conceptual details and implementation of these methods in SciPy.

Various methods of solving ODEs are as follows:

- Euler's method
- Taylor series method
- Runge-Kutta method
- Runge-Kutta fourth order formula
- Predictor-corrector method

The followings are some methods used to solve PDEs:

- Finite element method
- Finite difference method
- Finite volume method

The initial value problem

The initial value problem is an ordinary differential equation along with the value of an unknown function at a point in the solution domain, for example, $dy/dx = f(x,y)$, where, $y=y1$ for $x=x1$.

The boundary value problem

The boundary value problem is, again, a differential equation with some constraints, and its solution is the solution for the differential equation that satisfies these given constraints. These constraints are called boundary conditions.

Random number generator

In computation, the random number generator is an algorithm or process that generates a sequence of numbers that doesn't follow any pattern, which is why they are called random numbers. It is almost impossible to predict the number to be generated. The number of applications using random numbers is increasing day by day, and so it has led to the development of many methods for random number generation. This concept has been used for a long time, such as using dice, coin flipping, using playing cards, and many more methods. However, these methods have limited values for random numbers.

Computational methods of random number generation soon became popular for a wide variety of applications, such as statistical sampling, gambling, designing for random design generation, computerized simulation of various science and engineering concepts, and a number of other areas that demand unpredictable results, such as cryptography.

There are two main categories of random number generators, namely true random number generators and pseudo-random number generators. A true random number generator uses some physical phenomenon to generate a random number, for example, the actual read or write time taken by hard disk, whereas a pseudo-random number generator uses a computational algorithm for random number generation. There is also a third category of random number generators. They are based on statistical distributions, such as Poison distribution, exponential distribution, normal distribution, Gaussian distribution, and many more.

Various pseudo-random number generators are as follows:

- Blum Blum Shub
- Wichmann-Hill
- Complementary-multiply-with-carry
- Inversive congruential generator
- ISAAC (cipher)
- Lagged Fibonacci generator
- Linear congruential generator
- Linear-feedback shift register
- Maximal periodic reciprocals
- Mersenne twister
- Multiply-with-carry

- Naor-Reingold pseudo-random function
- Park–Miller random number generator
- Well-equidistributed long-period linear

Python scientific computing

Python's support for scientific computing is composed of a number of packages and APIs for different functionalities required for scientific computing. For each category, we have multiple options and a most popular choice. The following are the examples of Python scientific computing options:

- **Chart plotting**: At present, the most popular two-dimensional chart plotting package is matplotlib. There are several other plotting packages, such as Visvis, Plotly, HippoDraw, Chaco, MayaVI, Biggles, Pychart, and Bokeh. There are some packages that are built on top of matplotlib to provide enhanced functionality, such as Seaborn and Prettyplotlib.

- **Optimization**: The SciPy stack has an optimization package. The other choices for the optimization functionality are OpenOpt and CVXOpt.

- **Advanced data analysis**: Python supports integration with the R statistical package for advanced data analysis using RPy or the RSPlus-Python interface. There is a Python-based library for performing data analysis activities called pandas.

- **Database**: PyTables is a package for managing hierarchical databases. This package is developed on top of HDF5 and is designed to efficiently process large datasets.

- **Interactive command shell**: IPython is a Python package that supports interactive programming.

- **Symbolic computing**: Python has packages such as SymPy and PyDSTool for supporting symbolic computing. Later in this chapter, we are going to cover the idea of symbolic computing.

- **Specialized extensions**: SciKits provides special-purpose add-ons for SciPy, NumPy, and Python. The following a select list of Scikits packages:
  - `scikit-aero`: Aeronautical engineering calculations in Python
  - `scikit-bio`: Data structures, algorithms, and educational resources for bioinformatics
  - `scikit-commpy`: Digital communication algorithms with Python
  - `scikit-image`: Image processing routines for SciPy

- ○ scikit-learn: A set of Python modules for machine learning and data mining

- ○ scikit-monaco: Python modules for Monte Carlo integration

- ○ scikit-spectra: Spectroscopy in Python built on pandas

- ○ scikit-tensor: A Python module for multilinear algebra and tensor factorizations

- ○ scikit-tracker: Object detection and tracking for cell biology

- ○ scikit-xray: Data analysis tools for X-ray science

- ○ bvp_solver: A Python package for solving two-point boundary value problems

- ○ datasmooth: The Scikits data smoothing package

- ○ optimization: A Python module for numerical optimization

- ○ statsmodels: Statistical computations and models for use with SciPy

- **Third-party or non-scikit packages/applications/tools**: There are a number of projects that have developed packages/tools for specific fields of science, such as astronomy, astrophysics, bioinformatics, geosciences, and many more. The following are some selected third-party packages/tools in Python for specific scientific fields:

  - ○ Astropy: A community-driven Python package used to support astronomy and astrophysics computations

  - ○ Astroquery: This package is a collection of tools used to access online astronomy data

  - ○ BioPython: This is a collection of toolkits used to perform biological computations in Python

  - ○ HTSeq: This package supports the analysis of high-throughput sequencing data in Python

  - ○ Pygr: This is the toolkit for sequence and comparative genomic analysis in Python

  - ○ TAMO: This is a Python application used to analyze transcriptional regulation using DNA sequence motifs

  - ○ EarthPy: This is a collection of IPython notebooks that have examples from the earth science domain

  - ○ Pyearthquake: A Python package for earthquake and MODIS analysis

- ° `MSNoise`: This is a Python package for monitoring seismic velocity change using ambient seismic noise
- ° `AtmosphericChemistry`: This tool supports exploration, construction, and conversion of atmospheric chemistry mechanics
- ° `Chemlab`: This package is a complete library used to perform computations related to chemistry

Introduction to NumPy

Python programming is extended to support large arrays and matrices and a library of mathematical functions to manipulate these arrays. These arrays are multidimensional and this Python extension is called NumPy. After the success of the basic implementation of NumPy, it is extended with a number of APIs/ tools, including matplotlib, pandas, SciPy, and SymPy. Let's take a look at the brief functionality of each of the subtools/sub-APIs of NumPy.

The SciPy library

SciPy is Python library designed and developed for scientists and engineers for performing operations related to scientific computing. It supports functionalities for different operations, such as optimization, linear algebra, calculus, interpolation, image processing, fast Fourier transformation, signal processing, and special functions. It solves ODEs and performs other tasks required in science and engineering. It is built on top of the NumPy array object and is a very essential component of the NumPy stack. This is why the NumPy stack and the SciPy stack are sometimes used as the same reference.

The SciPy Subpackage

The various subpackages of SciPy include the following:

- `constant`: These are physical constants and conversion factors
- `cluster`: Hierarchical clustering, vector quantization, and K-means
- `fftpack`: Discrete Fourier transform algorithms
- `integrate`: Numerical integration routines
- `interpolate`: Interpolation tools
- `io`: Data input and output
- `lib`: Python wrappers to external libraries
- `linalg`: Linear algebra routines

- `misc`: Miscellaneous utilities (for example, image reading and writing)
- `ndimage`: Various functions for multidimensional image processing
- `optimize`: Optimization algorithms, including linear programming
- `signal`: Signal processing tools
- `sparse`: Sparse matrices and related algorithms
- `spatial`: KD-trees, nearest neighbors, and distance functions
- `special`: Special functions
- `stats`: Statistical functions
- `weave`: A tool for writing C/C++ code as Python multiline strings

Data analysis using pandas

The pandas library is an open source library designed to provide high-performance data manipulation and analysis functionalities in Python. Using pandas, users can process complete data analysis workflows in Python. Also, using pandas, the IPython toolkit, and other libraries, the Python environment for performing data analysis becomes very good in terms of performance and productivity. The pandas library has only one drawback; it supports only linear and panel regression. However, for other functionalities, we can use `statsmodels` and `scikit-learn`. The pandas library supports efficient merging and joining of datasets. It has bundles of tools for reading and writing data among different types of data sources, including in-memory, CSV, text files, Microsoft Excel, SQL databases, and the HDF5 format.

A brief idea of interactive programming using IPython

Python supports interactive computing in multiple programming languages with the help of IPython. IPython is a command shell especially designed for Python programming, and now it supports multiple languages. It offers excellent introspection functionality, new shell syntax, command-line text completion, and command history. Introspection is the capability of programming a command-line environment to examine various characteristics (properties, methods, and other details, such as the superclass). IPython has a number of features, including the following:

- Command-line-based and QT-based interactive shell
- A browser-based notebook that supports coding, mathematical expressions, inline graphics, and graphs

- It also has the capability to support interactive data visualization and other graphical user interfaces
- Support for high-performance parallel computing

IPython parallel computing

IPython has excellent support for parallel and distributed computing to facilitate large-scale computing. It has the capabilities for the development, execution, debugging, and monitoring of parallel or distributed applications. IPython supports most styles of parallelism, including the following, and any hybrid approach made from them:

- **Single program multiple data (SPMD)** parallelism
- **Multiple program multiple data (MIMD)** parallelism
- **Message Passing Interface (MPI)**
- Task and data parallelism
- Custom user-defined approaches

IPython Notebook

IPython Notebook is a Web-based interactive computation environment. This environment is used to create IPython notebooks. It takes single-user inputs or single expressions, evaluates them, and returns the result to the user. This functionality is called **read, evaluate, print, and looping (REPL)**. For REPL, the user can use the following Python libraries:

- IPython
- ØMQ (ZMQ)
- Tornado (web server)
- jQuery
- Bootstrap (frontend framework)
- MathJax

The notebook program creates a local web server on the computer to access it from a web browser. The IPython notebook is a JSON document used to perform different types of computations using coding, text, mathematical operations, graphics, and plotting. These notebooks can be exported to various formats using web-based and command-based options. The supported formats are HTML, LaTeX, PDF, Python, and many more.

The Python notebook development process is presented in the following figure. It starts from the left by preparing the data and then developing the program and its versioning. After the program development, it can be exported to various formats.

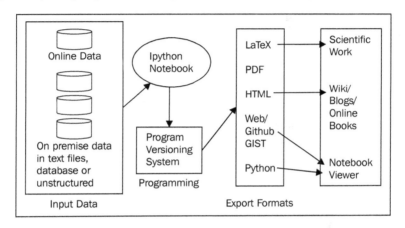

Some other remarkable features of IPython are as follows:

- Non-blocking interaction with GUI libraries and toolkits: IPython supports non-blocking interaction with a number of Python-based GUI toolkits/ libraries, including Tkinter, PyGTK, PyQt, and wxPython

- Cluster management: IPython supports computing the cluster management facility using MPI/asynchronous status callback messages

- Unix-like environment: The default behavior of IPython is almost similar to that of the Unix shells that support customization of the environment

Downloading the example code

You can download the example code files from your account at
http://www.packtpub.com for all the Packt Publishing books you
have purchased. If you purchased this book elsewhere, you can visit
http://www.packtpub.com/support and register to have the files
e-mailed directly to you.

A screenshot of the IPython user interface is depicted here (the source is http://ipython.org/notebook.html):

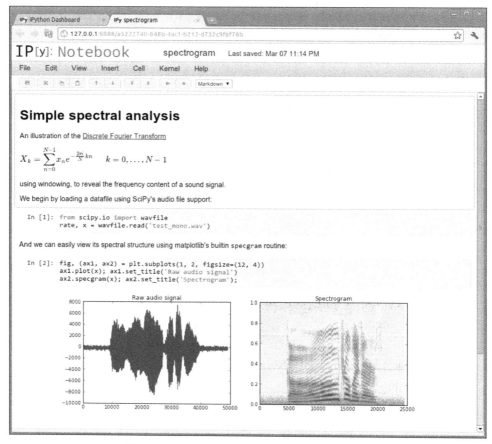

The user interface of the IPython Notebook interface

The following are various functional features of the IPython command shell:

- **Tab completion**: The user need not type complete commands. After typing only the initial few characters the remaining command can be completed by pressing *Tab*.

- **Exploring your objects**: Various properties of an object can be determined using the introspection facility.

- **Magic functions**: There are a number of magic functions that can be called by users.

- **Running and editing**: Users can execute and edit Python scripts from the command shell.

- **Debugging**: A strong debugging facility is also bundled with the command shell.

- **History**: The command shell stores the history of commands and their results.

- **System shell commands**: Users can also use the command provided by the system shell.

- **Define your own system aliases**: Users can define the aliases of the command as per their preferences.

- **Configuration**: IPython environment can be customized using the configuration files.

- **Startup files**: Users can customize the environment to run some commands or code at the beginning of the IPython session.

Symbolic computing using SymPy

Symbolic computation manipulates mathematical objects and expressions. These mathematical objects and expressions are represented as they are, and they are not evaluated/approximated. Expressions/objects with unevaluated variables are left in their symbolic form.

Let's see the difference between computerized normal computation and computerized symbolic computation in the following diagram. We have two examples each for both the cases. **Example A1** and **Example A2** are examples of normal computation, and **Example B1** and **Example B2** are examples of symbolic computation. **Example A1** and **Example A2** have obvious output. Let's take a look at the output of **Example B1** and **Example B2**. The output of **Example B1** is the same sqrt(3). No evaluation is performed; it's only the original symbols. This is because in symbolic computing, if the argument for the sqrt function is not a perfect square, then it will be left as it is. On the other hand, in **Example B2**, the output is slightly simpler. The reason is that for this example, it is possible to simplify the answer; sqrt(27) can be written as sqrt (9 X 3) or 3(sqrt(3), so it is simplified to 3sqrt(3).

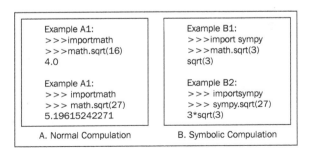

Comparison of normal computation and symbolic computation

The features of SymPy

As it is a symbolic computation library, SymPy has the ability to perform all types of computations symbolically. It can simplify expressions (as we have seen for sqrt(8)); compute differentiation, integration, and limits; and solve equations, matrix operations, and various other mathematical functions. All of these functionalities are performed symbolically.

Let's discuss the various features of SymPy. The SymPy library is composed of core capabilities and a number of optional modules. The following are the functionalities supported by SymPy:

- Core capabilities such as basic arithmetic and simplification, and pattern matching functions such as trigonometric, hyperbolic, exponential, logarithms, and many more
- It supports polynomial operations, for example, basic arithmetic, factorization, and various other operations
- The calculus functionality, for example, limits, differentiation, integration, and more
- Solving various types of equations, for example, polynomials, systems of equations, and differential equations
- Discrete mathematics
- The functionality of matrix representations and operations
- Geometric functions
- Plotting with the help of an external pyglet module
- Support for physics
- Performing statistical operations, such as probability and distributions
- Various printing functionalities
- Code generation for programming languages and LaTeX

Why SymPy?

SymPy is an open source library and is licensed under the liberal BSD license. You are allowed to modify the source code. This is not the case with other alternatives, such as Maple and Mathematica. Another advantage of SymPy is that it is designed, developed, and executed in Python. For a Python developer, this brings an added advantage. This library is highly extensible in comparison to alternative tools.

The plotting library

The chart plotting library of Python is named matplotlib. It provides an object-oriented API for addition of charts in an application developed using various Python GUI toolkits. SciPy/NumPy uses matplotlib to draw 2D charts of arrays. The design philosophy behind matplotlib is to simplify the plotting functionality. The user can easily create various types of plots using few function calls, or only one function call. There are some specialized toolkits/APIs that extend the functionality of matplotlib. Some of these tools are bundled with matplotlib, and others are available as separate downloads. Some of them are listed here:

- Basemap is a map plotting toolkit
- The Cartopy package is used to easily make drawing maps for data analysis and visualization
- Excel tools supports exchange of data with Microsoft Excel
- Interfaces for Qt and GTK+
- mplot3d can be used to draw 3D plots

The various types of the charts that can be plotted using matplotlib are given in the following table. These screenshots of charts have been taken from the matplotlib web page at `http://matplotlib.org/users/screenshots.html`:

Different types of graphs:	
Simple plot	Subplot demo with multiple axes

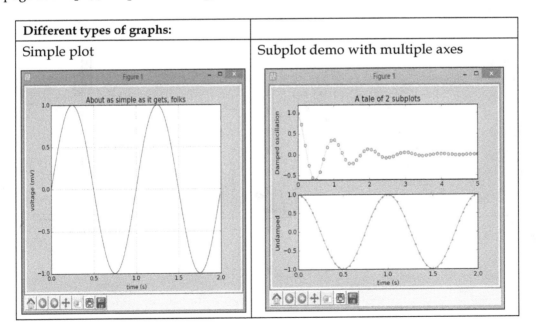

Histograms	Path demo
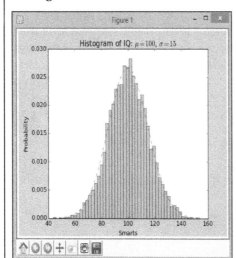	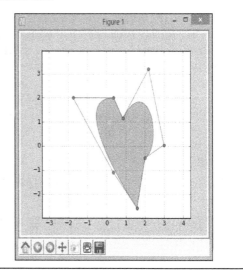

mplot3d: 3D Graphs	Stream plot: For plotting the streamlines of a vector field
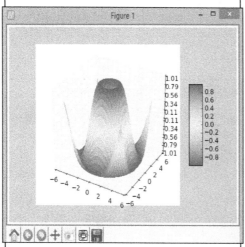	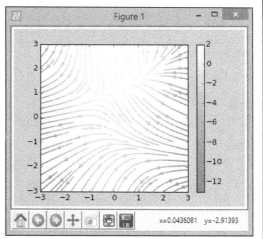

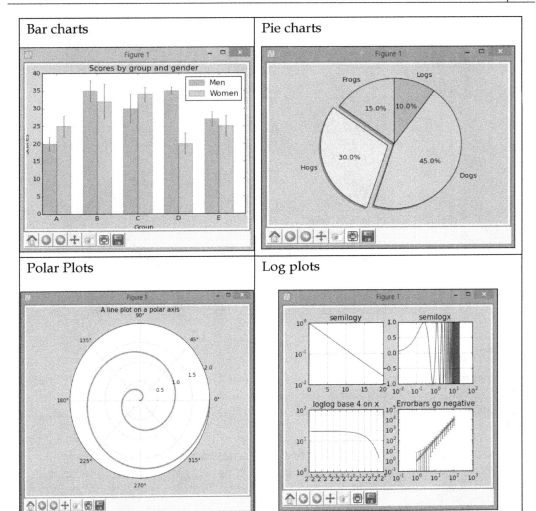

Financial charts: For drawing sophisticated financial plots by combining the various plot functions, layout commands, and labeling tools provided by matplotlib.

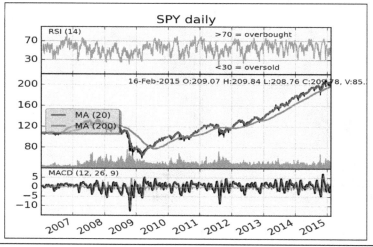

Summary

In this chapter, we discussed many concepts of mathematical and numerical analysis, including systems of linear and nonlinear equations, optimization, interpolation, extrapolation, numerical differentiation and integration, differential equations, and random number generators.

In the second part of the chapter, we briefly discussed the various packages/toolkits/APIs meant for performing scientific computing in the Python language. We also discussed the functionality and features of NumPy, SciPy, IPython, SymPy, matplotlib, and pandas.

In the next chapter, we will discuss how to prepare and manage data for scientific computations.

3
Efficiently Fabricating and Managing Scientific Data

This chapter is all about data for scientific computations. It introduces the concepts of this data, and then covers the various toolkits used to manage this data and the operations to be performed on it. After that, various data formats and random-number-based techniques for generating synthetic numerical data are discussed.

In this chapter, we will cover the following topics:

- The basics of data, information, and knowledge
- The concepts of various pieces of data storage software and tools
- Operations that can be performed on the data
- Details of the standard formats for scientific data
- A discussion on ready-to-use datasets
- Synthetic data generation using random numbers
- The idea of large-scale datasets

The basic concepts of data

A raw and unorganized form of facts and figures about an entity is called **data**. Any factual quantity or value in an unorganized/raw form (such as a series of numbers or alphabets) that represents a concept, phenomenon, object, or entity in the real world can be considered as data. There are no limits to data, and it is available everywhere.

Data can be transformed into information and can be useful for achieving the goals of its organization. There are certain properties that, when added to data, make it information. Accurate and timely data is called information if it is organized for a specific purpose and prepared and presented in a particular context. This gives a meaning and relevance to that data.

Data and information can further be transformed into knowledge by adding insights using domain experience. This knowledge requires vast experience of dealing with data related to the specific application, such as commodity price or weather forecasting.

Now, let's consider a scientific example of data, information, and knowledge. Obviously, 79 °F is a temperature reading and it is data. If we add some details along with this reading — for example, this is the temperature of the Gateway of India, Mumbai, India at 5:30 P.M. on March 3, 2015 — then it is information. On the basis of the hourly temperature readings of this particular week for a number of years, predicting the temperature of the next week is knowledge. Similarly, on the basis of some information on heavy snowfall in the last two days in northern India, concluding that the temperature of central India will also drop by certain degrees is knowledge. The relationship between data, information, and knowledge is depicted in the following figure. This process starts with data collected from experiments and then extracts information from this data. Finally, after a detailed analysis of this information, we interpret knowledge from it.

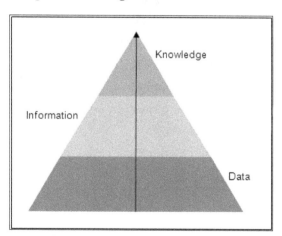

The pyramid of data, information, and knowledge

Data storage software and toolkits

Generally, the concepts involving computer science change very fast with time, and the software and tools for storing data evolve rapidly. Hence, at present, there are a number of pieces of software and toolkits available in the market for storing data. There are two major categories of data storage software and toolkits. Again, in each category, there are a number of subcategories. The taxonomy of various data-storing software pieces/toolkits used to manage and store data is depicted in this figure:

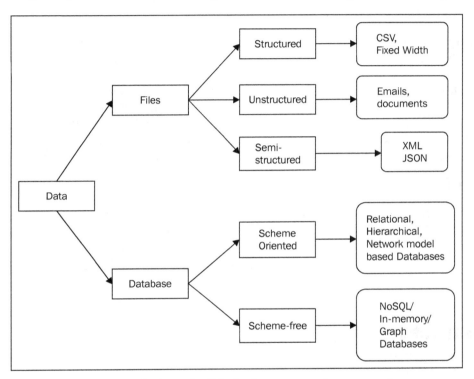

Taxonomy of software/toolkits for data storage and management

Files

The first category includes the software or tools that store data in flat files of different formats. The subcategories of flat files are structured and unstructured files. By "structured files," we mean files that have a predefined/fixed structure for storing data. Whereas, in unstructured files, there is no predefined structure to store the data. Generally, these two types of files store text data, while for some specific scientific applications, they may include images, audio, video, and other non-text data.

Structured files

An example of structured files may be text files with **Comma-separated Values (CSV)**. In such files, various data fields are separated by a comma or a delimiter. This delimiter can be any character or symbol. Preferably, this symbol must be a symbol that doesn't occur in the data to be stored. For example, if we are storing monetary values in our data, then a comma will not be a suitable choice for the delimiter.

Consider the following record of a CSV file: H.K. Mehta, 08-Oct-1975, Higher Education department, 50,432.

The preceding records have a name, date of birth, department name, and monthly salary. Now, for the CSV file, blank space, dot (.), comma (,), and dash (-) are not recommended as the delimiter for the fields. If we choose any one of these—blank space dot, comma, or dash—as the delimiter, then the comma will treat the amount as two values, and similarly, the dash (-) will treat the date of birth as three different values. The dot (.) will treat the name as three different values, and the department name will be divided into three values if blank space is the delimiter. For the records mentioned here, the delimiter can be one of other symbols, such as the question mark (?) or pipe (|). Generally, in commercial values, the pipe (|) is the most frequently used delimiter.

A fixed-width file is another example of a structured file. In such files, the total size of each field is predefined and maintained throughout the file. If the size for a field in a specific record is either smaller or larger than the predefined fixed size of that field, then for that particular record, either that field is padded with blank spaces (if it is smaller) or it is trimmed to reduce its size.

Unstructured files

Examples of unstructured files are web server logs, books, journals, and e-mails. This includes both text and non-text data. Text data includes data that can be represented by any character encoding scheme, such as **American Standard Code for Information Interchange (ASCII)** or Unicode. There is another category of file-based data stores, called semi-structured. It does not have the formal structure in accordance with the structure of relational and other databases. The semi-structured approach uses tags or other markers to separate fields, add appropriate meanings to the values, and create the structures of records and fields. Examples of such data are XML and JSON. The advantage of these data types is that they are of language and platform independent formats. Hence, their manipulation doesn't change with the language or platform.

Database

The products of the second category store data in a database. Besides files, there are a variety of databases for storing computational data. These databases may be divided into two main categories: schema-based databases and databases that have no schema. Schema-based databases are traditional databases that force the user to create structures before storing data. On the other hand, schema-free databases are a recent advancement in the field of large-scale databases, made to cope up with the demands of large-scale applications. Some examples of schema-based databases are MySQL, Oracle, MS SQL Server, Postgres and more. MongoDB, HBSE, and Cassandra are examples of no-schema databases.

Possible operations on data

Besides storing data, there are a number of operations that need to be performed to manage and use it effectively:

- **Data farming**: The process of using high-performance computing to run a set of simulations for a number of times on huge datasets is called data farming. The output of data farming is a vast view of the visible features and characteristics from the data; it supports the decision-making process. It is an integration of multiple disciplines, including high-performance computing, data analysis and visualization, and large-scale databases.

- **Data management**: Data management is a broad term that consists of a number of operations to be performed on the data, including the following:

    ◦ **Data governance**: This is the main control. It ensures that the entered data possesses the desired standards defined during its modeling. This data entry may be performed manually or by an automated process.

    ◦ **Data architecture, analysis, and design**: Data architecture involves applying various models, processes, algorithms, rules, or standards on what data has to be collected, how to structure the storing of the data, and how to integrate it. The analysis and design of the data involve the process of cleaning and transforming it to be useful for the benefit of the organization.

    ◦ **Database administration**: Administration of data is a complex task involving a number of related activities, such as development and designing of the database, database monitoring, overall system monitoring, improving the database's performance, database capacity and extension planning, security planning and implementation, and maintenance.

- ° **Data security management**: This includes management activities related to the security of the data, including access rights management, data privacy management, and other aspects related to the security of the data (such as data cleaning/wiping, encryption, masking, and backups).

- ° **Data quality management**: This is the task related to improving the quality of data. It involves a number of operations, including the following:

  - ° Cleansing of the data by detecting and correcting corrupt or inaccurate records of datasets, called dirty data.

  - ° Data integrity is the process of assuring data accuracy and consistency during different periods and processing on the data.

  - ° Data enrichment is the process of refining or enhancing the data with a focus on improving its quality by searching for misspellings or typographical errors according to the domains of the data. For example, in any condition, marks obtained cannot be more than the maximum possible marks. In such cases, we must correct those records that have marks greater than the maximum marks.

  - ° Data integration is a complicated process that requires extensive experience, as it combines data from a number of sources by converting them into a uniform structure, without affecting its meaning.

- ° **Data warehouse management**: Data warehouse management involves the process of preparation of a data mart, performing data mining, and performing various operations related to data movement, such as **extraction, translation, and loading** (ETL).

- ° **Metadata management**: This is the process of management of the data about the actual data stored in the database. This management data is called **metadata**. Metadata includes descriptions of the stored data, the date and time of data creation and modification, the owner of the data, its location on a physical device, and other related details.

- **Importing and exporting of data**: These are very important operations for any kind of application, whether it is a commercial application or a scientific application. Generally, care must be taken while performing import and export operations. The user needs to take into consideration the nature of the application for which they are exporting or importing the data. Accordingly, the proper format may be selected.

- **Scientific data archiving**: This is the process of storing scientific data for a long time based on the application and organizational policies about how much of the data the scientists should store and the accessibility level management of this data.

Scientific data format

There are a number of formats/forms available for storing scientific data. Generally, most scientific computation APIs/languages/toolkits support import and export operations on these formats. Some of the popular formats are as follows:

- **Network Common Data Form (NetCDF)**: This is a self-describing and machine/device/platform-independent data format that supports manipulation (creation, access, and sharing) of large array-based scientific data. It is also bundled with a set of software libraries for creation and manipulation. Generally, this format is used in applications such as weather forecasting, climate change from climatology and meteorology, the oceanography domain, and GIS applications. Most GIS applications support NetCDF as an input/output format, and it is also used for general scientific data exchanges.

  The main source of this format is the Unidata program at the **University Corporation for Atmospheric Research (UCAR)**. The project's home page is hosted by them. There is a popular quote from their website at `http://www.unidata.ucar.edu/software/netcdf/docs/faq.html#whatisit`:

  > *"NetCDF (network Common Data Form) is a set of interfaces for array-oriented data access and a freely-distributed collection of data access libraries for C, Fortran, C++, Java, and other languages. The NetCDF libraries support a machine-independent format for representing scientific data. Together, the interfaces, libraries, and format support the creation, access, and sharing of scientific data."*

- **Hierarchical Data Format (HDF)**: This is a set of file formats that have evolved in different versions (HDF4 and HDF5). This data format provides facilities to store and organize a large amount of numerical data. It was developed at the National Center for Supercomputing Applications, and now it is supported by the nonprofit HDF Group, which ensures further development of the HDF5 format and associated tools/technologies. This is a very popular format for scientific data, and HDF is supported by a number of tools, languages, technologies, and platforms, including Java APIs, MATLAB, Octave, Scilab, Python APIs, R APIs, and more.

- **Flexible Image Transport System (FITS)**: This is an open standard that defines a digital file format to manage image files used for scientific and other applications. This format is mainly used in astronomical applications. It has several options for describing photometric and spatial calibration information, along with other metadata. The first standardization of the FITS format was in 1981. Its most recent version was standardized in 2008. FITS can also be used to store non-image data, such as spectra, data cubes, or even databases. There is an important feature supported by FITS; the new version of the format is always kept backward compatible. Another important feature is that the metadata of the file is stored in human-readable ASCII characters in the header. This will help users analyze the file and understand the data stored in it.

- **Band-Interleaved Data/Band-Interleaved Files**: These are binary formats. This means that data is stored in non-text files, and generally, this format is used in remote sensing and high-end GIS. There are two subtypes of these files, namely **Band Interleaved by Line (BIL)** and **Band Interleaved by Pixel (BIP)**.

- **Common Data Format (CDF)**: This is a popular format for storing scalar and multidimensional platform-independent data. Hence, it is used to store scientific data and is popular as a data exchange format among researchers and organizations. The **Space Physics Data Facility (SPDF)** provides a CDF software toolkit as part of the **Goddard Space Flight Center (GSFC)** for manipulation of data. CDF also supports very good interfaces for a number of programming languages, tools, and APIs, including C, C++, C#, FORTRAN, Python, Perl, Java, MATLAB, and IDL.

There are a number of common features of various scientific data formats, as follows:

- These formats support both sequential-access and random-access reading of data.

- They are designed to efficiently store large volumes of scientific data.

- These formats contain the metadata for supporting the self-description capability.

- These formats, by default, support ordering of objects, grids, images, and ndarrays.

- These formats are not updatable. The user can append data at the end.

- They support machine portability.

- Most of these formats are standardized.

The various data formats discussed here can be used to store data of any subject and its subdomain. However, there are certain data formats specially designed for particular subjects. The following is a list of subject-specific data formats. I have not given any description for these subject-specific formats; readers with specific interests may refer to their origin:

- Formats for astronomical data:
    - **FITS**: The FITS astronomical data and image format (`.fit` or `.fits`)
    - **SP3**: GPS and other satellite orbits (`.sp3`)

- Format for storing medical imaging data:
    - **DICOM**: DICOM-annotated medical images (`.dcm`, `.dic`)

- Formats for medical and physiological data:
    - **Affymetrix**: The Affymetrix data format (`.cdf`, `.cel`, `.chp`, `.gin`, `.psi`)
    - **BDF**: The BioSemi data format (`.bdf`)
    - **EDF**: The European data format (`.edf`)

- Formats for chemical and biomolecular data:
    - MOL
    - SDF
    - SMILES
    - PDB
    - GenBank
    - FASTA

- Format for seismographic (earthquake-related science and engineering) data:
    - **NDK**: The NDK seismographic data format (`.ndk`)

- Format for weather data:
    - **GRIB**: The GRIB scientific data format (`.grb`, `.grib`)

Ready-to-use standard datasets

Several government, collaborative, and research efforts are continuously going on to develop and maintain standard datasets for different subjects and domains inside subjects. These datasets are available for the public to download or to work offline, or they can also have the facility of online computations over these datasets. One such notable effort is named **Open Science Data Cloud (OSDC)**, which has several datasets on each subject. This list, compiled from various open data sources, is available. They also host data on their web portal (`https://www.opensciencedatacloud.org/publicdata/`). A subject-wise list of selected datasets from OSDC is as follows:

- Agriculture:
    - The U.S. Department of Agriculture's plants database

- Biology:
    - 1,000 genomes
    - Gene Expression Omnibus (GEO)
    - MIT cancer genomics data
    - Protein data bank

- Climate/weather:
    - Australian weather
    - Canadian Meteorological Centre
    - Climate data from UEA (updated monthly)
    - Global climate data Since 1929

- Complex networks:
    - CrossRef DOI URLs
    - The DBLP citation dataset
    - NIST complex network data collection
    - UFL sparse matrix collection
    - The WSU graph database

- Computer networks:
    - 3.5 B web pages from Common Crawl 2012
    - 53.5 B web clicks of 100,000 users in Indiana University
    - CAIDA Internet datasets
    - ClueWeb09—1B web pages

- Data challenges:
  - ° Challenges in machine learning
  - ° DrivenData competitions for social good
  - ° The ICWSM Data Challenge (since 2009)
  - ° Kaggle competition data

- Economics:
  - ° American Economic Association (AEA)
  - ° EconData from UMD
  - ° Internet product code database

- Energy:
  - ° AMPds
  - ° BLUEd
  - ° Dataport
  - ° UK-Dale

- Finance:
  - ° CBOE futures exchange
  - ° Google Finance
  - ° Google Trends
  - ° NASDAQ

- "Geospace"/GIS:
  - ° BODC—marine data of about 22,000 vars
  - ° Cambridge, MA, US, GIS data on GitHub
  - ° EOSDIS—NASA's earth-observing system data
  - ° Geospatial data from ASU

- Healthcare:
  - ° EHDP large health datasets
  - ° Gapminder World—demographic databases
  - ° Medicare Coverage Database (MCD), USA
  - ° The Medicare data file

- Image processing:
    - 2 GB of photos of cats
    - Affective image classification
    - Face recognition benchmark
    - Massive Visual Memory Stimuli, MIT
    - The SUN database, MIT

- Machine learning:
    - Discogs monthly data
    - eBay online auctions (2012)
    - The IMDb database
    - The Keel repository for classification, regression, and time series
    - The Million Song Dataset

- Museums:
    - Cooper-Hewitt's collection database
    - Minneapolis Institute of Arts metadata
    - Tate Collection metadata
    - The Getty vocabularies

- Natural language:
    - Blogger Corpus
    - ClueWeb09 FACC
    - Google Books Ngrams (2.2 TB)
    - Google Web 5gram (1 TB in 2006)

- Physics:
    - The CERN open data portal
    - NSSDC (NASA) data of 550 spacecraft

- Public domain:
    - The CMU JASA data archive
    - The UCLA SOCR data collection
    - UFO reports
    - WikiLeaks 911 pager intercepts

- Search engines:
    - ° Academic torrents of data sharing from UMB
    - ° Archive-it from Internet Archive
    - ° DataMarket (Qlik)
    - ° Statista.com— statistics and Studies

- Social sciences:
    - ° The CMU Enron Email dataset of 150 users
    - ° Facebook social network from LAW (since 2007)
    - ° The Foursquare social network in 2010 and 2011
    - ° Foursquare from UMN/Sarwat (2013)

- Sports:
    - ° Betfair historical exchange data
    - ° Cricsheet matches (baseball)
    - ° Ergast Formula 1, from 1950 to present day (API)
    - ° Football/soccer resources (data and APIs)

- Time series:
    - ° Time Series Data Library (TSDL) from MU
    - ° UC Riverside time series dataset
    - ° Hard drive failure rates

- Transportation:
    - ° Airlines' OD data from 1987 to 2008
    - ° Bike Share Systems (BSS) collection
    - ° Bay Area bike share data
    - ° Hubway Million Rides in MA
    - ° Marine traffic—ship tracks, port calls, and more

Data generation

For some applications, if the user does not have data that can be used for computations, then they need to generate that data before performing computations. It can be generated in three ways: it can be collected personally, collected by instruments, or (for some specific applications) generated synthetically on computers.

There are some applications for which data is supposed to be collected personally; for example, if an application requires biometric data of a person, the data may be collected personally by setting up a data collection and requesting volunteers to support the biometric data collection. This collection must be performed personally, as this data cannot be produced on computers or using instruments. For this specific application, there is a possibility that the users get support from the government in order to obtain such data from governmental databases, such as the databases of biometric details collected during visa processing, or a nationwide project such as person registration databases of USA government or data collected during the unique identification project (ADHAAR) in India.

For some specific experiments, the data can be generated using a number of instruments that provide the readings of the users interested. For example, weather-related data can be generated using instruments as follows: we can place a number of temperature recorders at different places and periodically collect their readings. Using some specialized sensors, we can also collect weather- or health-science-related data. For example, the pulse rate and blood-pleasure-related information can be collected using specially designed smart belts or watches distributed to various persons, and the information can be collected from a built-in GPS system within these devices that will periodically use the push or pull method.

Synthetic data can be generated for a number of experiments that require numerical or text data, as these are pieces of data that may be produced on computers without any specific instruments, using a program that generates the data as per the predefined constraints. To generate text data, existing offline text data or online web pages that have text information may be used to generate new samples for processing. For example, text mining and linguistic processing sometimes require sample text data.

Synthetic data generation (fabrication)

In this section, we will discuss the various methods of synthetic numerical data generation. We will also present an algorithm for random number generation using the Poisson distribution and its Python implementation. Furthermore, we will explore different methods for synthetic text data generation.

Using Python's built-in functions for random number generation

Python has a module named random that implements various pseudo-random number generators on the basis of various statistical distributions. This module has functions for various types of randomness, such as for integers, for sequences, for random permutations of a list, and to generate a random sample from a predefined population. The Python random module supports random number generation using various statistical distributions, including uniform, normal (Gaussian), lognormal, negative exponential, gamma, and beta distributions. To generate a uniform random angle, Python provides the von Mises distribution. Most of the modules of the random number generator in Python depend on a basic function named random(). This function generates random floating-point numbers in a semi-open range (*[0.0, 1.0)*).

Mersenne Twister is the main random number generator of Python. It is capable of producing random floating point numbers of 53-bit precision with the period of 2**19937-1. It is written on top of the underlying implementation in C, which is thread-safe and fast. This is one of the most extensively used and tested random number generators. However, it is not suitable for all applications, as it is completely deterministic. Hence, it is not at all appropriate for security-related computations. The random module also provides a SystemRandom class, which generates random numbers using the os.urandom() function from the facility provided by the operating system. This class can be used to generate random numbers to be used for cryptographic purposes.

The functions of the random module are bound methods of a hidden instance of the random.Random class. However, the user can have their own instance of the Random class. The advantage is that this instance doesn't share the state. Moreover, if the user requires designing of a new random number generator, then this class can also be extended/inherited to create a new subclass of Random. In this situation, the user is supposed to override five methods: getstate(), jumpahead(), random(), seed(), and setState().

Let's discuss the various built-in methods of the Python random module. These functions are divided into categories, as discussed in the following section.

Bookkeeping functions

Various bookkeeping functions of the random module are as follows:

- `random.seed(a=None, version=2)`: This function initializes the random number generator. If the user passes an integer value to `a`, then that value is used. If no value is passed to `a` or if it is `none`, then the present system time is used as the `seed` value. If the operating system being used supports the randomness sources, then they will be used instead of the system time as a `seed` value.

- `random.getstate()`: This function returns an object that represents the current internal state of the random number generator. This object can be used to restore the the same state using the `setstate()` function.

- `random.setstate(state)`: The state value must be the object obtained from a call to the `getstate()` function. Then, `setstate` will restore the internal state of the generator to the state that it was having when `getstate()` was called.

- `random.getrandbits(k)`: This function returns a Python long integer with `k` random bits. This method is provided with the `MersenneTwister` generator and optionally by a few other generators.

Functions for integer random number generation

Different functions that return integer random numbers are stated here:

- `random.randrange(stop)` or `random.randrange(start, stop[, step])`: This method returns a randomly selected element from the given. The meanings of its parameters are as follows: `start` is the starting point of the range; it will be included in the range. The `stop` function is the terminating point of the range; it will be excluded from the range. The `step` represents the value to be added to a number to decide a random number.

- `random.randint(a,b)`: This function returns an integer value within the inclusive range from `a` to `b`.

Functions for sequences

Various functions that operate on sequences to generate a new random sequence or subsequence are as follows:

- `random.choice(seq)`: This function returns a random element of the non-empty `seq` sequence. The `seq` character must be non-empty. If `seq` is empty, then this function raises an error/exception called `IndexError`.

- `random.shuffle(x)`: This function shuffles the x sequence in place. Shuffling in place means that the positions of the values will be changed inside of the list variable.

- `random.sample(population, k)`: This function returns a k length list of unique random elements from the population. The population must be a sequence or set. This function is generally used for random sampling without replacement. Moreover, the members of the population may be duplicated, and each of their occurrences has equal probability of being present in the selected list. A `ValueError` exception will be raised if the size of sample is larger than the population size *k*.

Statistical-distribution-based functions

There are many statistical distributions suitable for various cases. In order to support these cases, the random number module has a bundle of functions for different statistical distributions. The following are the statistical-distribution-based random number generators:

- **Random number generator function (random.uniform(a, b))**: This function returns a random floating-point number *N* between the range of a and b. There is equal probability of selecting any number between a and b.

- **The random.triangular(low, high, mode) generator**: This function returns a random floating-point number *N* as per the triangular distribution such that *low <= N <= high*. The low and high values are treated as bound, and the mode is kept between these bounds. The default value of the low bound is 0. It is 1 for high bounds, and the mode argument defaults to the midpoint between the low and high bounds.

- **The random.betavariate(alpha, beta) generator**: This function returns a random number between 0 and 1 as per the beta distribution conditions on the parameters as *alpha(a) > 0* and *beta (β)> 0*.

- **The random.expovariate(lambd) generator**: This function returns a random number according to the exponential distribution. The value of the argument lambd (⊠) should be nonzero. It returns values in the range of 0 to positive infinity if lambd is positive, and from negative infinity to 0 if lambd is negative. This argument is intentionally termed lambd as lambda is a reserve word in Python.

- **The random.gammavariate(alpha, beta) generator**: This function generates random numbers by following the gamma distribution, with the conditions on the parameters as *alpha (a) > 0* and *beta (β)> 0*.

- **The random.normalvariate(mu, sigma) generator**: The normal distribution is followed to generate random numbers. Here, mu (μ) is the mean and sigma (σ) is the standard deviation.

- **The random.gauss(mu, sigma) generator**: As the name suggests, the Gaussian distribution is used in this function to generate random numbers. Again, mu is the mean and sigma is the standard deviation. In comparison to the normal distribution, this function is faster.

- **The random.lognormvariate(mu, sigma) generator**: The log normal distribution is used for random number generation. The natural logarithm of the values obtained by this distribution gives the value of the normal distribution. Once again, mu is the mean and sigma is the standard deviation. Here, mu can have any value, but sigma must be greater than zero.

- **The random.vonmisesvariate(mu, kappa) generator**: This function returns random angles using the von Mises distribution, where mu is the mean angle in radians, with a value between 0 and *2*pi*, and kappa (κ) is the concentration parameter (>=0).

- **The random.paretovariate(alpha) generator**: This function follows the Pareto distribution to return a random variable. Here, alpha is the shape parameter.

- **The random.weibullvariate(alpha, beta) generator**: The Weibull distribution is used in this function to generate random numbers. Here, alpha is the scale parameter and beta is the shape parameter.

Nondeterministic random number generator

Besides the discussed random number generation functions, there is an alternative random number generator available that can be used especially for those situations when random number generation must be nondeterministic, such as numbers required in cryptography and security. This generator is as class `random.SystemRandom([seed])`. This class generates random numbers using the `os.urandom()` function provided by the operating system.

The following program demonstrates the use of the functions discussed. The output of the function call is also presented in it. For simplicity, we have only used the following functions in it:

- `random.random`
- `random.uniform`
- `random.randrange`
- `random.choice`

- `random.shuffle`
- `print random.sample`
- `random.choice`

The program is as follows:

```
import random
print random.random()
print random.uniform(1,9)
print random.randrange(20)
print random.randrange(0, 99, 3)
print random.choice('ABCDEFGHIJKLMNOPQRSTUVWXYZ') # Output 'P'
items = [1, 2, 3, 4, 5, 6, 7, 8, 9, 10]
random.shuffle(items)
print items
print random.sample([1, 2, 3, 4, 5, 6, 7, 8, 9, 10],  5)
weighted_choices = [('Three', 3), ('Two', 2), ('One', 1), ('Four', 4)]
population = [val for val, cnt in weighted_choices for i in
range(cnt)]
print random.choice(population)
```

Let's discuss the output of each function call. The first function, `random`, returns any floating-point random values greater than 0 and less than 1. The `uniform` function returns uniformly distributed random values between the given ranges.
The `randrange` function returns a random integer value in the given range. If the first argument is ignored, then it will take its default value, 0. So, for `randrange(20)`, the range is 0 to 19.

Now, let's discuss the output of the functions related to the sequences. The `choice` function returns a random choice from the list of choices provided. In this example, there are 26 choices, and one value, P, is returned. The output of the `shuffle` function is obvious, and as expected, some of the values are shuffled. The `sample` function selects a random sample of a given size. In this example, the `sample` size is selected as 5. Hence, the random sample has five elements. The last three lines perform an important functionality of selecting a random choice with the given probability. That's why this `choice` function is called a weighted choice — as the weight is assigned to each of the choices required by the applications.

Designing and implementing random number generators based on statistical distributions

In this section, we will be discussing the designing of an algorithm and its Python implementation for Poisson distribution. This will be good in two aspects; one is that you will learn about the design and implementation of a new statistical distribution for random number generation. The second aspect is that this function is not available in the random module, so users can also use this new distribution. For some specific applications, some of the variables assume Poisson random values. For example, consider the scheduling algorithms used in process scheduling in operating systems. To simulate process scheduling, the process arrival follows the Poisson distribution.

There are a number of situations where the Poisson distribution is applied. Some of these cases are as follows:

- The pattern of traffic on the Internet follows a Poisson distribution
- The number of calls received at a call center follows Poisson distribution
- The number of goals made in game such as hockey or football (with two teams) also follows Poisson distribution
- The process arrival time in operating systems
- Given an age group, the number of deaths in a year is, again, a Poisson pattern
- The number of jumps in a stock price in a given time interval
- If we apply radiations to a given stretch of DNA, the number of mutations follows Poisson distribution

The algorithm of Poisson distribution is given by Knuth in his popular book *The Art of Computer Programming, Volume 2*, and it is as follows:

```
algorithm poisson_random_number (Knuth):
    initializations:
        L = e-⊠,
count = 0
product = 1
    do:
            k = k + 1
            u = uniform_random_number (0,1)
p = p × u
    while p > L
    return k - 1
```

The following code is the Python implementation of the Poisson distribution:

```python
import math
import random
def nextPoisson(lambdaValue):
  elambda = math.exp(-1*lambdaValue)
  product = 1
  count = 0

  while (product >= elambda):
    product *= random.random()
    result = count
    count+=1
  return result
for x in range(1, 9):
  print nextPoisson(8)
```

The output of the preceding program is as follows:

```
5
7
11
8
9
8
7
6
```

A special note on reproducing the random number generated

If an application demands reproduction of the random number generated using any method, then in such cases, there is an option for reproducing the numbers generated using these functions. To reproduce the sequence, we just need to use the same function with the same seed value. In this way, we can reproduce the list, and this is why we call most random number generation functions deterministic.

A program with simple logic to generate five-digit random numbers

The next program demonstrates the idea of using time and date objects to produce random numbers. It has very simple logic for generating five-digit random numbers. In this program, the current system time is used to generate the random numbers. The four components of the system time—hours, minutes, seconds, and microseconds—are generally a unique combination. This value is converted to a string and then to a five-digit value. The first line in the user-defined function is used to introduce a delay at the microsecond level so that the time value will be different among different calls within very short time. Without this line, the user may get some repeated values:

```
import datetime
import time

# the user defined function that returns 5 digit random number
def  next_5digit_int():
  # this will introduce randomness at the microsecond level
  time.sleep(0.123)
current_time = datetime.datetime.now().time()
  random_no = int(current_time.strftime('%S%f'))
  # this will trim last three zeros
  return random_no/1000

# to demonstrate generation of ten random numbers
for x in range(0, 10):
  i = next_5digit_int()
  print i
```

A brief note about large-scale datasets

The datasets of various scientific applications range from several MB to a few GB. For some specific applications, the datasets may be huge. These gigantic datasets may span up to a couple of petabytes. We usually understand MB and GB; let's just get an idea of the scale of a petabyte. Suppose we store one petabyte of data in **compact disks (CDs)** and arrange these CDs in the form of a stack. The size of this stack will be approximately 1.75 kilometers. Due to recent advances in networking and distributed computing technologies, these days, there are a number of applications that process datasets of several petabytes. In order to efficiently process large-scale datasets, there are a number of options available at all levels of software or hardware.

There are several efficient frameworks for processing datasets of all scales. These frameworks can process small-, medium-, or large-scale data with equal efficiency, depending on the infrastructure provided. Map reduce is an example of such a framework, and Hadoop is an open source implementation of the MapReduce framework.

At the database level, the user has a number of choices that are capable of storing and managing data of any scale. These databases may be the simplest ones like flat files — either text or binary. Then, there are a number of schema-based databases, such as relational databases, that can efficiently manage a database of several gigabytes. Both files and schema-based databases can manage data from megabytes to several gigabytes. To process data beyond these limits, the trend these days is to use non-schema-based databases and advanced distributed filesystems, for example, Google's BigTable, Apache HBase, and HDFS. HBase is a column-oriented database designed to support very-large-scale databases. HDFS is a distributed filesystem that is capable of storing files of size of several petabytes, unlike the maximum file size of around 16 GB in a normal filesystem, such as WINDOWS NT.

Most programming languages support these frameworks and databases, including Python, Scala, Java, Ruby, and more. Besides the software level, there are advances at the hardware level as well, such as the concept of virtualization in different pieces of hardware (for example, processors, I/O devices, and networking devices). There is also enhancement of hardware-level support for the software discussed.

A recent advancement in distributed computing, called cloud computing, has enabled a number of new scientific computing and commercial applications. This is possible because cloud computing, together with the concepts discussed in this section, provides the highest ever processing and storage power. This has enabled a number of new applications, and the list of such applications is growing day by day.

The technologies discussed are extensively used in applications that require text searching, pattern finding and matching, image processing, data mining, and crunching of huge datasets. Such requirements are very frequent in various commercial and scientific applications.

In *Chapter 8, Parallel and Large-scale Scientific Computing*, we will have detailed a discussion of these technologies, with a focus on using them for large-scale scientific applications.

Summary

This chapter began with a discussion of the basic concepts of data, information, and knowledge. Then it introduced the various types of software used to store data. After that, we discussed various operations that should be performed on datasets. Then we saw the standard format of storing scientific data. We also discussed various predefined, already used, and standard datasets for a number of scientific applications in various subject domains. However, there are some domains in particular subjects for which datasets not be available.

After covering the basic concepts, various techniques of preparation of synthetic data for some specific experiments were presented. Various standard functions available for random number generation used in synthetic data generation were also presented. For synthetic data generation, one algorithm and a program for random number generation using the Poisson distribution were covered.

The next chapter will have a detailed discussion on, and show the functionality of, various Python APIs and toolkits for scientific computing. These APIs provide numerical computation (NumPy and SciPy), symbolic computing (SymPy), data visualization and plotting (matplotlib and pandas), and interactive programming (IPython). The chapter will also present a brief discussion of the features and functionalities of these APIs.

Scientific Computing APIs for Python

In this chapter, we will have a comprehensive discussion on the features and capabilities of the various scientific computing APIs and toolkits in Python. Besides the basics, we will also discuss some example programs for each of the APIs. As symbolic computing is a relatively different area of computerized mathematics, we have allocated a special subsection within the SymPy section to discuss the basics of the computerized algebra system.

In this chapter, we will cover the following topics:

- Scientific numerical computing using NumPy and SciPy
- Symbolic computing using SymPy
- A computerized algebra system
- Introduction to SymPy and its modules
- A few simple exemplary programs in SymPy
- Data analysis, visualization, and interactive computing

Numerical scientific computing in Python

Scientific computing mainly demands the facility of performing calculations on algebraic equations, matrices, differentiations, integrations, differential equations, statistics, equation solvers, and more. By default, Python doesn't come with these functionalities. However, the development of NumPy and SciPy has enabled us to perform these operations and far more advanced functionalities beyond them. NumPy and SciPy are very powerful Python packages that enable users to efficiently perform the desired operations for all types of scientific applications.

The NumPy package

NumPy is the basic Python package for scientific computing. It provides the facilities of multidimensional arrays and basic mathematical operations, such as linear algebra. Python provides several data structures to store user data; the most popular data structures are lists and dictionaries. List objects may store any type of Python object as an element. These elements can be processed using loops or iterators. Dictionary objects store data in the key value format.

The ndarrays data structure

ndarrays are similar to list but are highly flexible and efficient. An ndarray is an array object used to represent multidimensional arrays of fixed-size items. This array should be homogeneous. It has an associated object of the type `dtype` for defining the data type of the elements in the array. This object defines the type of data (integer, float, or Python object), the size of data in bytes, and the byte ordering (big-endian or little-endian). Moreover, if the type of data is `record` or `sub-array`, then it also contains details about them. The actual array can be constructed using any one of the array, zeros, or empty methods.

Another important aspect of ndarrays is that the size of arrays can be dynamically modified. Moreover, if the user needs to remove some elements from the arrays, then this can be done using the module for masked arrays. In a number of situations, scientific computing demands deletion/removal of incorrect or erroneous data. The `numpy.ma` module provides the facility of the masked array to easily remove selected elements from arrays. A masked array is nothing but the normal ndarrays with a mask. Mask is another associated array with true or false values. If, for a particular position, mask has a true value, then the corresponding element in the main array is valid, and if the mask is false, then the corresponding element in the main array is invalid or masked. In such a case where the value is `false`, while performing any computation on such ndarrays, the masked elements will not be considered.

File handling

Another important aspect of scientific computing is storing data in files, and NumPy supports reading and writing on both text as well as binary files. Mostly, text files are a good way of reading, writing, and data exchange as they are inherently portable and most platforms, by default, have the capability to manipulate them. However, for some applications, it is sometimes better to use binary files, or in some cases the desired data for such application can be stored in binary files only. Sometimes, the size and nature of data such as an image or a sound require it to be stored in binary files.

In comparison with text files, binary files are harder to manage as they have specific formats. Moreover, the size of binary files is comparatively very small and read/write operations for them are much faster than those for read/write text files. This fast read/write is most suitable for applications working on large datasets. The only drawback of binary files manipulated with NumPy is that they are accessible only through NumPy.

Python has text file manipulation functions, such as `open`, `readlines`, and `writelines`. However, it is not performance efficient to use these functions for scientific data manipulation. These default Python functions are very slow in reading and writing data in a file. NumPy has a high-performance alternative that loads data into ndarrays before the actual computation. In NumPy, text files can be accessed using the `numpy.loadtxt` and `numpy.savetxt` functions. The `loadtxt` function can be used to load data from text files to ndarrays. NumPy also has a separate function for manipulating data in binary files. The functions for reading and writing are `numpy.load` and `numpy.save`, respectively.

Some sample NumPy programs

The NumPy array can be created from a list or tuple that uses the array. This method can transform sequences of sequences into two-dimensional arrays:

```
import numpy as np
x = np.array([4,432,21], int)
print x    #Output [  4 432  21]
x2d = np.array( ((100,200,300), (111,222,333), (123,456,789)) )
print x2d
```

Here is the output:

```
[  4 432  21]
[[100 200 300]
[111 222 333]
[123 456 789]]
```

Basic matrix arithmetic operations can easily be performed on two-dimensional arrays, as used in the following program. Basically, these operations are applied on elements. Hence, the operand arrays must be of equal size. If the sizes do not match, then performing these operations will cause a runtime error. Consider the following example for arithmetic operations on one-dimensional arrays:

```
import numpy as np
x = np.array([4,5,6])
y = np.array([1,2,3])
print x + y    # output [5 7 9]
```

```
print x * y     # output [ 4 10 18]
print x - y     # output [3 3 3]
print x / y     # output [4 2 2]
print x % y     # output [0 1 0]
```

There is a separate subclass named matrix for performing matrix operations. Let's understand matrix operations by the following example, which demonstrates the difference between array-based multiplication and matrix multiplication. NumPy matrices are two-dimensional and arrays can be of any dimension:

```
import numpy as np
x1 = np.array( ((1,2,3), (1,2,3), (1,2,3)) )
x2 = np.array( ((1,2,3), (1,2,3), (1,2,3)) )
print "First 2-D Array: x1"
print x1
print "Second 2-D Array: x2"
print x2
print "Array Multiplication"
print x1*x2

mx1 = np.matrix( ((1,2,3), (1,2,3), (1,2,3)) )
mx2 = np.matrix( ((1,2,3), (1,2,3), (1,2,3)) )
print "Matrix Multiplication"
print mx1*mx2
```

The output is as follows:

```
First 2-D Array: x1
[[1 2 3]
 [1 2 3]
 [1 2 3]]
Second 2-D Array: x2
[[1 2 3]
 [1 2 3]
 [1 2 3]]
Array Multiplication
[[1 4 9]
 [1 4 9]
 [1 4 9]]
Matrix Multiplication
[[ 6 12 18]
 [ 6 12 18]
 [ 6 12 18]]
```

The following is a simple program that demonstrates simple statistical functions given in NumPy:

```
import numpy as np
x = np.random.randn(10)        # Creates an array of 10 random elements
print x
mean = x.mean()
print mean
std = x.std()
print std
var = x.var()
print var
```

This is the first sample output:

```
[0.08291261  0.89369115  0.641396   -0.97868652  0.46692439 -
  0.13954144
 -0.29892453  0.96177167  0.09975071  0.35832954]
0.208762357623
0.559388806817
0.312915837192
```

The following is the second sample output:

```
[ 1.28239629  0.07953693 -0.88112438 -2.37757502  1.31752476
  1.50047537
  0.19905071 -0.48867481  0.26767073  2.660184  ]
0.355946458357
1.35007701045
1.82270793415
```

The preceding programs are some simple examples of NumPy. In *Chapter 5*, *Performing Numerical Computing* we will have a detailed discussion on the NumPy functionality. The next subsection discusses the SciPy Python package.

The SciPy package

SciPy extends Python and NumPy support by providing advanced mathematical functions, such as differentiation, integration, differential equations, optimization, interpolation, advanced statistical functions, equation solvers, and many more. SciPy is written on top of the NumPy array framework. It has utilized the arrays provided in NumPy and the basic operations on them, and has extended it to cover most of the mathematical aspects that are regularly required by scientists and engineers for their applications.

In this chapter, we will cover examples of some basic functionality, and in *Chapter 5, Performing Numerical Computing*, we will have exhaustive coverage of the NumPy and SciPy functionalities. In subsequent subsections, we will cover the basics of the various important packages/modules of SciPy, including clustering analysis, file handling, integration, interpolation, optimization, signal and image processing, special analysis, and statistics.

The optimization package

The optimization package in SciPy provides the functionality to solve univariate and multivariate minimization problems. It provides solutions for minimization problems using a number of algorithms and methods. The minimization problem has a wide range of applications in science and commercial domains. Generally, we perform linear regression, searching for a function's minimum and maximum values, finding the root of a function, and linear programming for such cases. All of these functionalities are supported by the optimization package.

The interpolation package

A number of interpolation methods and algorithms are provided in the interpolation package as built-in functions. It provides the facility to perform univariate and multivariate interpolation and one-dimensional and two-dimensional splines. We use univariate interpolation when data is dependent on one variable; if it is dependent on more than one variable, then we use multivariate interpolation. Besides these functionalities, the interpolation package also provides additional functionality for Lagrange and Taylor polynomial interpolators.

Integration and differential equations in SciPy

Integration is an important mathematical tool for scientific computations. The SciPy integrations subpackage provides functionalities to perform numerical integration. SciPy provides a range of functions to perform integration on equations and data. It also has an ordinary differential equation integrator. It provides various functions to perform numerical integrations with the help of a number of methods from mathematics using numerical analysis.

The stats module

The SciPy stats module contains a function for most probability distributions and wide-range or statistical functions. Supported probability distributions include various continuous distributions, multivariate distributions, and discrete distributions. The statistical functions range from simple means to the most complex statistical concepts, including skewness, kurtosis, and the chi-square test, to name a few.

Clustering package and spatial algorithms in SciPy

Clustering analysis is a popular data mining technique that has a wide range of applications in scientific and commercial domains. In the science domain, biology, particle physics, astronomy, life science, and bioinformatics are a few subjects that widely use clustering analysis for problem solving. Clustering analysis is used extensively in computer science for computerized fraud detection, security analysis, image processing, and many more areas. The clustering package provides functionality for K-means clustering, vector quantization, and hierarchical and agglomerative clustering functions.

The `spatial` class has functions for analyzing the distance between data points using triangulations, Voronoi diagrams, and convex hulls of a set of points. It also has KDTree implementations for performing the nearest-neighbor lookup functionality.

Image processing in SciPy

SciPy provides support for performing various image processing operations, including basic reading and writing of image files, displaying images, and simple image manipulation operations such as cropping, flipping, and rotating. It also has support for image filtering functions, such as mathematical morphing, smoothing, denoising, and sharpening of images. Furthermore, it supports various other operations, such as image segmentation by labeling pixels corresponding to different objects, classification, and feature extraction for example edge detection.

Sample SciPy programs

In subsequent subsections, we will discuss some example programs that use SciPy modules and packages. We will start with a simple program that performs standard statistical computations. After that, we will discuss a program that finds a minimal solution using optimizations. Finally, we will discuss image-processing programs.

Statistics using SciPy

The stats module of SciPy has functions to perform simple statistical operations and various probability distributions. The following program demonstrates simple statistical calculations using the `stats.describe` SciPy function. This single function operates on an array and returns the number of elements, minimum value, maximum value, mean, variance, skewness, and kurtosis:

```
import scipy as sp
import scipy.stats as st
s = sp.randn(10)
```

```
n, min_max, mean, var, skew, kurt = st.describe(s)
print("Number of elements: {0:d}".format(n))
print("Minimum: {0:3.5f} Maximum: {1:2.5f}".format(min_max[0],
    min_max[1]))
print("Mean: {0:3.5f}".format(mean))
print("Variance: {0:3.5f}".format(var))
print("Skewness : {0:3.5f}".format(skew))
print("Kurtosis: {0:3.5f}".format(kurt))
```

Here is the output:

```
Number of elements: 10
Minimum: -2.00080 Maximum: 0.91390
Mean: -0.55638
Variance: 0.93120
Skewness : 0.16958
Kurtosis: -1.15542
```

Optimization in SciPy

Generally, in mathematical optimization, a non-convex function called the Rosenbrock function is used to test the performance of the optimization algorithm. The following program demonstrates the minimization problem on this function. The Rosenbrock function of N variables is given by the following equation, and it has a minimum value 0 at $xi = 1$:

$$f(x) = \sum_{i=1}^{N-1} 100\left(x_i - x_{i-1}^2\right)^2 + \left(1 - x_{i-1}\right)^2$$

The program for the preceding function is as follows:

```
import numpy as np
from scipy.optimize import minimize

# Definition of Rosenbrock function
def rosenbrock(x):
    return sum(100.0*(x[1:]-x[:-1]**2.0)**2.0 + (1-x[:-1])**2.0)

x0 = np.array([1, 0.7, 0.8, 2.9, 1.1])
res = minimize(rosenbrock, x0, method = 'nelder-mead', options =
{'xtol': 1e-8, 'disp': True})

print(res.x)
```

This is the output:

```
Optimization terminated successfully.
        Current function value: 0.000000
        Iterations: 516
        Function evaluations: 827
[ 1.  1.  1.  1.  1.]
```

The last line is the output of `print(res.x)`, wherein all the elements of the array are 1.

Image processing using SciPy

The following two programs have been developed to demonstrate the image processing functionality of SciPy. The first of these programs simply displays a standard test image. This image is widely used in the field of image processing and is called Lena. The second program applies geometric transformation on this image. It performs image cropping and rotation by 45 percent.

The following program displays the Lena image using the matplotlib API. The `imshow` method renders the ndarrays into an image, and the `show` method displays the image:

```
from scipy import misc
l = misc.lena()
misc.imsave('lena.png', l)
import matplotlib.pyplot as plt
plt.gray()
plt.imshow(l)
plt.show()
```

The output of the previous program is shown in the following screenshot:

The following program performs geometric transformation. This program displays the transformed images along with the original image as a four-axis array:

```
import scipy
from scipy import ndimage
import matplotlib.pyplot as plt
import numpy as np

lena = scipy.misc.lena()
lx, ly = lena.shape
crop_lena = lena[lx/4:-lx/4, ly/4:-ly/4]
crop_eyes_lena = lena[lx/2:-lx/2.2, ly/2.1:-ly/3.2]
rotate_lena = ndimage.rotate(lena, 45)

# Four axes, returned as a 2-d array
f, axarr = plt.subplots(2, 2)
axarr[0, 0].imshow(lena, cmap=plt.cm.gray)
axarr[0, 0].axis('off')
axarr[0, 0].set_title('Original Lena Image')
axarr[0, 1].imshow(crop_lena, cmap=plt.cm.gray)
axarr[0, 1].axis('off')
axarr[0, 1].set_title('Cropped Lena')
axarr[1, 0].imshow(crop_eyes_lena, cmap=plt.cm.gray)
axarr[1, 0].axis('off')
axarr[1, 0].set_title('Lena Cropped Eyes')
axarr[1, 1].imshow(rotate_lena, cmap=plt.cm.gray)
axarr[1, 1].axis('off')
axarr[1, 1].set_title('45 Degree Rotated Lena')

plt.show()
```

This is the output:

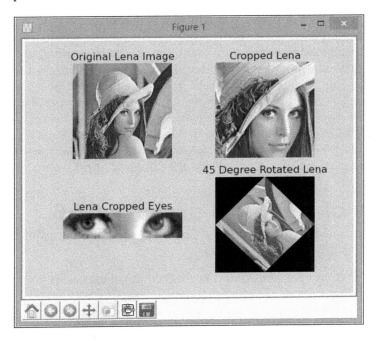

SciPy and NumPy are the core of Python's support for scientific computing, as they provide solid functionality in numerical computing. In *Chapter 5, Performing Numerical Computing*, we will be discussing both the packages in detail. The next subsection introduces symbolic computing using SymPy.

Symbolic computations using SymPy

Computerized computations performed on mathematical symbols without evaluating or changing their meaning are called symbolic computations. Generally, symbolic computing is also called computerized algebra, and such a computerized system is called a computer algebra system. The following subsection has a brief and good introduction to SymPy. In *Chapter 6, Applying Python for Symbolic Computing*, we will have in-depth coverage of symbolic computing in Python.

Computer Algebra System

Let's discuss the concept of **Computer Algebra System (CAS)**. CAS is a piece of software or toolkit used to perform computations on mathematical expressions using computers instead of computing manually. In the beginning, using computers for these applications was called computer algebra; now this concept is called symbolic computing. CAS systems may be grouped into two types. The first type is general-purpose CAS, and the second type is CAS specific to a particular problem. General-purpose systems are applicable to most areas of algebraic mathematics, while specialized CAS are systems designed for specific areas, such as group theory or number theory. Most of the time, we prefer general-purpose CAS for manipulation of mathematical expressions for scientific applications.

Features of a general-purpose CAS

Various desired features of a general-purpose CAS for scientific applications are as follows:

- A user interface for manipulating mathematical expressions.

- An interface for programming and debugging.

- Such systems require simplification of various mathematical expressions. Hence, a simplifier is the most essential component of this type of computerized algebra system.

- A general-purpose CAS system must support an exhaustive set of functions to perform the various mathematical operations required by any algebraic computation.

- Most applications perform extensive computations, so efficient memory management is highly essential.

- The system must provide support for performing mathematical computations on high-precision numbers and large quantities.

A brief idea of SymPy

SymPy is an open source and Python-based implementation of CAS. The philosophy behind the development of SymPy is to design and develop a CAS that has all the desired features and yet whose code is as simple as possible so that it will be highly and easily extensible. It is written completely in Python and does not require any external library.

The basic idea behind SymPy is the creation and manipulation of expressions. Using SymPy, the user represents mathematical expressions in the Python language—by using SymPy classes and objects. These expressions are composed of numbers, symbols, operators, functions, and more. The functions are the modules to perform a mathematical functionality, such as logarithms and trigonometry.

The development of SymPy was started by Ondřej Čertík in August 2006. Since then, it has grown considerably with contributions from hundreds of people. This library now consists of 26 different integrated modules. These modules have the capability to perform computations required for basic symbolic arithmetic, calculus, algebra, discrete mathematics, quantum physics, plotting, and printing, along with the option to export the output of the computations in LaTeX and other formats.

The capabilities of SymPy can be divided into two categories—*core capability* and *advanced capabilities*—as the SymPy library is divided into a core module and several advanced optional modules. The functionalities supported by various modules are discussed in the following sections.

Core capability

The core capability module supports the basic functionalities required by any mathematical algebra operation to be performed. These operations include basic arithmetic, such as multiplication, addition, subtraction, and division, and also exponentials. The module also supports simplification of expressions in order to simplify complex expressions. It provides the functionality of expansion of series and symbols.

The core module also supports functions meant to perform operations related to trigonometry, hyperbolas, exponentials, roots of equations, polynomials, factorials, gamma functions, logarithms, and a number of special functions for B-Splines, spherical harmonics, tensor functions, and orthogonal polynomials.

There is also strong support for pattern matching operations in the core module. Furthermore, the core capabilities of SymPy include functionalities to support the substitutions required by algebraic operations. It supports not only high-precision arithmetic operations over integers, rational numbers, and floating-point numbers, but also non-commutative variables and symbols required in polynomial operations.

Polynomials

Various functions used to perform polynomial operations belong to the polynomial module. These functions include division, **greatest common divisor (GCD)**, **least common multiplier (LCM)**, square-free factorization, representation of polynomials with symbolic coefficients, some special operations such as the computation of the resultant, deriving trigonometric identities, partial fraction decomposition, and facilities for the Gröbner basis over polynomial rings and fields.

Calculus

Various functionalities that support the different operations required by basic and advanced calculus are provided in the calculus module. It supports functionalities required by limits; there is a `limit` function for this. It also supports differentiation, integration, series expansion, differential equations, and calculus of finite differences. SymPy also has special support for definite integrals and integral transforms. In differentials, it supports numerical differentials, composition of derivatives, and fractional derivatives.

Solving equations

Solver is the name of the SymPy module that provides the equation solving functionality. This module supports solving capabilities for complex polynomials, roots of polynomials, and systems of polynomial equations. There is a function for solving algebraic equations. It not only provides support for solutions of differential equations (including ordinary differential equations, some forms of partial differential equations, initial and boundary values problems, and more), but also supports solutions of difference equations. In mathematics, a difference equation is also called a recurrence relation, that is, an equation that recursively defines a sequence or multidimensional array of values.

Discrete math

Discrete mathematics includes mathematical structures that are discrete in nature rather than continuous mathematics (such as calculus). It deals with integers, graphs, and statements from the logic theory. This module has full support for binomial coefficients, products, and summations.

This module also supports various functions from the number theory, including the residual theory, Euler's Totient, partition, and a number of functions dealing with prime numbers and their factorizations. Plus SymPy supports the creation and manipulation of logic expressions using symbolic and Boolean values.

Matrices

SymPy has strong support for various operations related to matrices and determinants. Matrices belong to the linear algebra category of mathematics. It supports the creation of matrices, basic matrix operations (such as multiplication and addition), matrix of zeros and ones, creation of a random matrix, and performing operations on matrix elements. It also supports special functions, such as computation of the Hessian matrix for a function, the Gram-Schmidt process on a set of vectors, computation of the Wronskian for a matrix of functions, and more.

Furthermore, it has full support for eigenvalues and eigenvectors, matrix inversion, and solutions of matrices and determinants. To compute the determinants of a matrix, it supports Bareis' fraction-free algorithm and Berkowitz's algorithm, besides other methods. For matrices, it supports null space calculation, cofactor expansion tools, derivative calculation for matrix elements, and calculating the dual of a matrix.

Geometry

SymPy has a module that supports various operations associated with 2D geometry. It supports the creation of 2D entities or objects such as a point, line, circle, ellipse, polygon, triangle, ray, and segment. It also allows us to perform queries on these entities, such as the area of a suitable object (ellipse, circle, or triangle) and the intersection points of lines. Then, it supports queries such as line tangency determination and finding the similarity and intersection of entities.

Plotting

There is a very good module that allows us to draw two-dimensional and three-dimensional plots. At present, plots are rendered using the `matplotlib` package. It also supports other packages, such as `TextBackend`, `Pyglet`, `textplot`, and more. It has a very good interactive interface facility of customizations and plotting of various geometric entities.

The plotting module has functions for plotting the following:

- 2D line plots
- 2D parametric plots
- 2D implicit and region plots
- 3D plots of functions involving two variables
- 3D line and surface plots

Physics

There is also a module for solving problems from the physics domain. It supports functionality for mechanics, including classical and quantum mechanics, and High-energy Physics. It has functions that support Pauli algebra and quantum harmonic oscillators in one dimension and three dimensions. It also has functionality for optics. There is a separate module that integrates unit systems into SymPy. This allows users to select the specific unit system for performing their computations and conversions between units. The unit systems are composed of units and constants for computations.

Statistics

The statistics module was introduced in SymPy to support the various concepts of statistics that are required in mathematical computations. Apart from supporting various continuous and discrete statistical distributions, it also supports functionality related to symbolic probability. Generally, these distributions support functions for random number generation in SymPy.

Printing

SymPy has a module for providing full support for *Pretty-Printing*. Pretty-printing converts various kind of stylistic formatting into text files such as source code, text files, markup files or similar content. This module produces the desired output by printing using ASCII and/or Unicode characters.

It supports various printers, such as LaTeX and the MathML printer. It is capable of producing source code in various programming languages, such as C, Python, and Fortran. It is also capable of producing content using markup languages such as HTML and XML.

SymPy modules

The following list shows the formal names of the modules discussed in preceding paragraphs:

- **Assumptions**: The assumption engine
- **Concrete**: Symbolic products and summations
- **Core basic class structure**: Basic, Add, Mul, Pow, and so on
- **Functions**: Elementary and special functions
- **Galgebra**: Geometric algebra
- **Geometry**: Geometric entities

- **Integrals**: Symbolic integrator
- **Interactive**: Interactive sessions (for example, IPython)
- **Logic**: Boolean algebra and theorem proving
- **Matrices**: Linear algebra and matrices
- **mpmath**: Fast arbitrary precision numerical math
- **ntheory**: Number theoretical functions
- **Parsing**: Mathematica and maxima parsers
- **Physics**: Physical units and quantum stuff
- **Plotting**: 2D and 3D plots using `Pyglet`
- **Polys**: Polynomial algebra and factorization
- **Printing**: Pretty-printing and code generation
- **Series**: Symbolic limits and truncated series
- **Simplify**: Rewriting expressions in other forms
- **Solvers**: Algebraic, recurrence, and differential
- **Statistics**: Standard probability distributions
- **Utilities**: Test frameworks and compatibility-related content

There are numerous symbolic computing systems available in various mathematical toolkits. There are some pieces of proprietary software, such as Maple and Mathematica, and there are some open source alternatives as well, such as Singular and AXIOM. However, these products have their own scripting language. It is difficult to extend their functionality, and they have slow development cycles. On the other hand, SymPy is highly extensible, is designed and developed in the Python language, and is an open source API that supports a speedy development life cycle.

Simple exemplary programs

Here are some very simple examples to help you get an idea about the capacities of SymPy. These are fewer than 10 lines of SymPy source code each, and they cover topics ranging from basic symbol manipulations to limits, differentiation, and integration. We can test the execution of these programs on SymPy by live-running SymPy online on Google App Engine, available at http://live.sympy.org/.

Basic symbol manipulation

The following code defines three symbols and an expression with these symbols. Then it prints the expression:

```
import sympy
a = sympy.Symbol('a')
b = sympy.Symbol('b')
c = sympy.Symbol('c')
e = ( a * b * b + 2 * b * a * b) + (a * a + c * c)
print e
```

The output is as follows:

```
a**2 + 3*a*b**2 + c**2
```

Here, ** represents a power operation.

Expression expansion in SymPy

The program shown here demonstrates the concept of expression expansion. It defines two symbols and a simple expression on these symbols and then prints the expression and its expanded form:

```
import sympy
a = sympy.Symbol('a')
b = sympy.Symbol('b')
e = (a + b) ** 4
print e
print e.expand()
```

This is the output:

```
(a + b)**4
a**4 + 4*a**3*b + 6*a**2*b**2 + 4*a*b**3 + b**4
```

Simplification of an expression or formula

SymPy has the facility to simplify mathematical expressions. The following program has two expressions for simplifying, and it displays the output after simplifications of the expressions:

```
import sympy
x = sympy.Symbol('x')
a = 1/x + (x*exp(x) - 1)/x
simplify(a)
simplify((x ** 3 +  x ** 2 - x - 1)/(x ** 2 + 2 * x + 1))
```

Here is the output:

```
ex
x - 1
```

Simple integrations

This program calculates the integration of two simple functions:

```
import sympy
from sympy import integrate
x = sympy.Symbol('x')
integrate(x ** 3 + 2 * x ** 2 + x, x)
integrate(x / (x ** 2 + 2 * x), x)
```

The output is as follows:

```
x**4/4+2*x**3/3+x**2/2
log(x + 2)
```

APIs and toolkits for data analysis and visualization

Python has excellent toolkits and APIs that are used to analyze, visualize, and present data and the results of computations. In the subsequent discussion, we will cover the concept and idea of pandas. We will briefly discuss matplotlib and some sample programs on chart drawing and exporting in different formats. We can export charts in image files and other files, such as PDF. In *Chapter 7, Data Analysis and Visualization* we will have a detailed discussion on most of the concepts of matplotlib and pandas, along with the IPython toolkits.

Data analysis and manipulation using pandas

pandas is a Python package for data analysis and data manipulation. It is composed of a number of data structures for working on scientific data analysis in Python. The ultimate goal behind the development of pandas is to design a powerful and flexible data manipulation and analysis tool. It provides efficient, flexible, and significant data structures, specially designed so that they can work with any kind of data. pandas can be used to work with most types of popular databases and datasets. pandas is developed on top of NumPy.

Hence, it inherently supports integration with the other scientific computing APIs and toolkits of Python. It can be used with any of the following types of data:

- It can be tabular data, such as relational databases or spreadsheets (for example, MS Excel)

- It may be ordered or unordered time series data

- It may be data organized in multidimensional arrays, such as a matrix with row and column labels

- It may be any dataset used to store scientific data in any formats that we discussed in *Chapter 3, Efficiently Fabricating and Managing Scientific Data*

Important data structures of pandas

The pandas data structures range from 1D up to 3D. Series is 1D, DataFrame is 2D, and panel is the three-dimensional and higher dimensional data structure; its higher dimension such as 4D is under development. Generally, series and DataFrame are suitable for most use cases in statistics, engineering, finance, and social science:

- **Series**: This is a labeled 1D array that may be used to store any data type, such as integers, floating-point numbers, strings, and other valid Python objects. The labels of this axis are collectively referred to as the index.

- **DataFrame**: This is a labeled 2D data structure with rows and columns. The columns may have different types. DataFrame may be considered similar to other 2D structures such as spreadsheet tables and database tables. DataFrame can also be considered as a collection of multiple series of different types.

- **Panel**: In statistics and economics, panel data refers to multidimensional data that contains different measurements taken over time. The name of this data structure is derived from this concept. In comparison to series and DataFrame, panel is a less used data structure.

Special features of pandas

The following are the highlighting features of pandas:

- It provides the facility of data manipulation between the pandas data structures in the memory and different data formats, including CSV, Microsoft Excel, SQL databases, and the HDF5 format.

- It is highly optimized for achieving high performance; the critical code is developed in Cython and C.

- It supports the partition or subdivision of large datasets using slicing, indexing, and subsetting.

- It provides automatic and explicit data alignment. Objects can be aligned explicitly to a set of labels. If the user ignores the labels, then the data structures automatically align the data.

- The data structures support dynamic size mutability, as columns can be inserted and deleted.

- pandas has a powerful engine for group by operations used in aggregation and transformation of data.

- It also supports efficient merge and join operations on datasets for data integration.

- It uses the concept of reindexing to manage missing data. By "missing data," we mean null or absent data.

- pandas also has excellent support for time-series-specific functionalities, including moving windows statistics, date range generation and frequency conversion, date shifting and lagging, moving window linear regressions, and much more.

Data visualization using matplotlib

matplotlib is the Python API meant for data visualization. It is the most widely utilized Python package for 2D graphics. It provides a fast and customizable way of data visualization and publication of quality images in a number of formats. It supports drawing of multidimensional charts. matplotlib has default values for most of the properties of these charts. However, these values are highly customizable. The user can control almost all the settings of any chart, such as figure size, line width, color and style, axes, axis and grid properties, and text properties (such as font, face, and size).

Let's discuss some examples of drawing and exporting them into different formats.

Interactive computing in Python using IPython

There are two popular styles of working with Python programs: either interactively or through scripts. There are still some programmers who prefer to work with scripts. Generally, they use a text editor to write their programs, and use the terminal for execution and other activities, such as debugging. Also, scientific computing applications generally demand a very good interactive computing environment. In interactive computing, the processes may take input from humans whenever required. This input may be taken from the command line or a graphic user interface. Python scientific computing APIs get an interactive computing environment using the set of tools bundled with IPython. IPython is heavily used in various activities in scientific computing applications, such as data management, data manipulation, data analysis, data visualization, scientific computations, and large-scale computations.

Let's discuss some simple examples of using IPython for computations by using NumPy, SymPy, pandas, and matplotlib.

Sample data analysis and visualization programs

In this subsection, we will discuss sample programs for data analysis and visualization using matplotlib and pandas. You can use live IPython available at https://www.pythonanywhere.com/try-ipython/ if you don't have a local installation of pandas and matplotlib.

To begin with, we need some data to analyze or visualize. The following program fetches data about Apple from Yahoo! finance from October 1, 2014 to January 31, 2015, and saves this data in a CSV file:

```
import pandas as pd
import datetime
import pandas.io.data

start = datetime.datetime(2014, 10, 1)
end = datetime.datetime(2015, 1, 31)

apple = pd.io.data.get_data_yahoo('AAPL', start, end)
print(apple.head())
apple.to_csv('apple-data.csv')
df = pd.read_csv('apple-data.csv', index_col='Date', parse_dates=True)
df.head()
```

This is the output:

	Open	**High**	**Low**	**Close**	**Volume**	**Adj close**
Date						
10/1/2014	100.59	100.69	98.7	99.18	51491300	98.36
10/2/2014	99.27	100.22	98.04	99.9	47757800	99.08
10/3/2014	99.44	100.21	99.04	99.62	43469600	98.8
10/6/2014	99.95	100.65	99.42	99.62	37051200	98.8
10/7/2014	99.43	100.12	98.73	98.75	42094200	97.94

The next program makes a plot of the data from the `.csv` file created in the previous example. It calculates 50 moving averages (50 MA) on the close reading. Then it plots the open, close, high, low, and 50 moving average data in a 2D plot. The chart shown in this screenshot is prepared by the program shown after the screenshot:

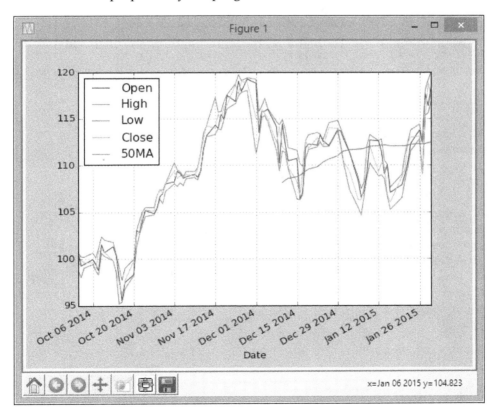

Here's the program:

```
import pandas as pd
import matplotlib.pyplot as plt

df = pd.read_csv('apple-data.csv', index_col = 'Date', parse_
dates=True)
df['H-L'] = df.High - df.Low
df['50MA'] = pd.rolling_mean(df['Close'], 50)
df[['Open','High','Low','Close','50MA']].plot()
plt.show()
```

Now, the following program makes a 3D plot of the same data:

```
import pandas as pd
import matplotlib.pyplot as plt
from mpl_toolkits.mplot3d import Axes3D

df = pd.read_csv('apple-data.csv', parse_dates=True)
print(df.head())
df['H-L'] = df.High - df.Low
df['50MA'] = pd.rolling_mean(df['Close'], 50)

threedee = plt.figure().gca(projection='3d')
threedee.scatter(df.index, df['H-L'], df['Close'])
threedee.set_xlabel('Index')
threedee.set_ylabel('H-L')
threedee.set_zlabel('Close')
plt.show()

threedee = plt.figure().gca(projection='3d')
threedee.scatter(df.index, df['H-L'], df['Volume'])
threedee.set_xlabel('Index')
threedee.set_ylabel('H-L')
threedee.set_zlabel('Volume')
plt.show()
```

The output of the preceding program is a 3D plot, as shown in the following screenshot:

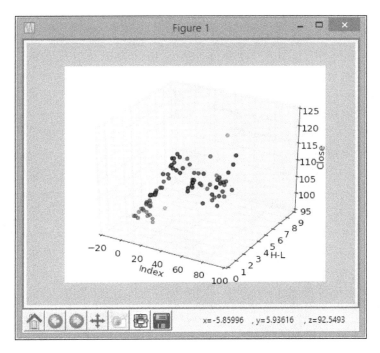

Summary

In this chapter, we discussed the concepts, the features, and some selected sample programs of various scientific computing APIs and toolkits. The chapter started with a discussion of NumPy and SciPy. After covering NymPy, we discussed the concepts associated with symbolic computing and SymPy.

In the remaining chapter, we discussed interactive computing and data analysis and visualization, along with their APIs or toolkits. IPython is the Python toolkit for interactive computing. We also discussed the data analysis package named pandas and the data visualization API named matplotlib.

In next chapter, we will have a detailed discussion on the numerical computing API—NumPy. We will cover various functions of NumPy and the associated mathematical concepts with the help of sample programs.

5
Performing Numerical Computing

In this chapter, we will discuss most of the features of NumPy and SciPy with the help of example programs. We will start with a detailed discussion on arrays and the operations that can be performed on them, using examples. This will lay a solid foundation for discussing various advanced functionalities supported by NumPy and SciPy.

In this chapter, we will cover the following topics:

- Scientific numerical computing using NumPy and SciPy
- The fundamental objects of NumPy
- The various packages/modules of NumPy
- The basics of the SciPy package
- Mathematical functions of SciPy
- Advanced mathematical modules and packages

NumPy is the base of numerical computing in Python, and its most fundamental and important idea is support for multidimensional arrays. Let's start our discussion with the underlying concepts of arrays in NumPy. After the basics, we will discuss the various operations that can be performed on multidimensional arrays. We will also cover the various basic and advanced mathematical functions supported by NumPy. NumPy has some subpackages or modules for supporting advanced mathematical concepts.

The NumPy fundamental objects

The entire scientific computing functionality of NumPy and SciPy is built around two basic types of objects in NumPy. The first object is an *n*-dimensional array object known as **ndarray**, and the second object is a universal function object called `ufunc`. Besides these two objects, there are a number of other objects built on top of them.

The ndarray object

The ndarray object is a homogenous collection of elements that are indexed using *N* integers, where *N* is the dimension of the array. There are two important attributes of ndarray. The first is the data type of the elements of the array, called `dtype`, and the second is the shape of the array. The data type here can be any data type supported by Python. The shape of the arrays is an *N*-tuple, that is, a collection of *N* elements for the *N*-dimensional array, where each element of the tuple defines the number of elements in that dimension of the array.

The attributes of an array

Besides the shape and `dtype`, the other important attributes of an array are as follows:

- `size`
- `itemsize`
- `data`
- `ndim`

The `itemsize` is the length of one element of the array in bytes, and the `data` attribute is a Python buffer object that points to the start of the array's data. Let's understand the concept of shape, data type, and other attributes with the help of the following Python program:

```
import numpy as np
x2d = np.array( (  (100,200,300),
                   (111,222,333),
                   (123,456,789) ) )
x2d.shape
x2d.dtype
x2d.size
x2d.itemsize
x2d.ndim
x2d.data
```

Basic operations on arrays

The use of square brackets ([]) to index array values is known as **array indexing**. Consider the x2d two-dimensional array defined and used in the previous program. A particular element of a two-dimensional array may be referred to as x2d[row, column]. For example, we can refer to the second element of the second row (that is, 222) as x2d[1,1], as the index starts from 0. Similarly, the x2d[2,1] element means the second element of the third row (that is, 456).

Array slicing is the process of selecting some elements from an array to produce a subarray. For a single-dimensional array, we can sequentially select some elements from the array. Further, by using slicing , we can fetch an entire row or column of a two-dimensional array. In other words, using slicing, we can fetch array elements across an axis. The basic slicing concept of Python is extended to *N* dimensions for ndarray. The basic slice syntax is start:stop:step. The first element specifies the start index of the slice, the second element specifies the stop index of the slice, and the last element defines the step to be added to the index of the previously selected element. If we skip any of the first two values, then that value is considered zero or more. Similarly, the default value of step is 1. Let's consider a few examples to make slicing more clear.

Consider x2d, a 6 x 3 array. Then, x2d[1] is the same as x2d[1, :] and represents the second row of the array, which has three elements. On the other hand, x2d[:, 1] represents the second column of the array, which has six elements. Every third element of the second column can be selected as x2d[:: 1, 2].

Ellipses can be also be used to replace zero or more : terms. An ellipsis expands to zero or more full-slice objects to match the total dimensions of the sliced object, which is equal to the dimensions of the original array. For example, if x4d is 5×6×7×8, then x4d[2 :, ..., 6] is equivalent to x4d[2 :, :, :, 6]. Similarly, x4d[..., 4] is equivalent to A[:, :, :, 4]. Consider the following program to get a clear idea of the concept of array slicing. This program demonstrates the slicing of one-dimensional and two-dimensional arrays:

```
import numpy as np
x = np.array([1,12, 25, 8, 15, 35, 50, 7, 2, 10])
x[3:7]
x[1:9:2]
x[0:9:3]

x2d = np.array((  (100,200,300),
                  (111,222,333),
                  (123,456,789),
```

```
                      (125,457,791),
                      (127,459,793),
                      (129,461,795) ))
    x2d[0:4,0:3]
    x2d[0:4:2,0:3:2]
```

The iteration over an array can be performed using a `for` loop. In one-dimensional arrays, we can fetch all the elements sequentially using a `for` loop. On the other hand, iterating over multidimensional arrays can be performed with respect to the first axis. This program demonstrates how to perform iterations over arrays:

```
import numpy as np
x = np.array([1,12, 25, 8, 15, 35, 50, 7, 2, 10])
x2d = np.array((  (100,200,300),
                (111,222,333),
                (123,456,789),
                (125,457,791),
                (127,459,793),
                (129,461,795) ))
for i in x:
  print i

for row in x2d:
  print row
```

Special operations on arrays (shape change and conversion)

For changing the shape of an array, we have many methods such as `ravel`, `reshape`, `resize`, and assigning new value to the shape attribute. The `ravel` and `reshape` methods return the argument (the calling object) with the modified shape, while the resizing and assignment modify the actual array. The ravel method flattens the array into a C-language-style array. It returns the argument as if it is a one-dimensional array, with each row sequentially arranged one by one.

Let's discuss the impact of these methods with the help of the next program. This program performs shape manipulation operations on a two-dimensional array. The first `print` in the program will display the original array. The `ravel` function will display the flattened array. The `print` function after the `ravel` function will display the original array again, as the `ravel` function doesn't change the original array. Now, the `resize` function will change the shape of the array from the original shape (6,3), which has six rows and three columns, to (3,6), which has three rows and six columns. Hence, the `print` function after the `resize` function will display the array in its new shape.

Now, we have applied the `reshape` function on the original shape of the array
(`(6,3)`). This will display the array with the original shape of `(6,3)`. However, as
`reshape` doesn't change the shape, the `print` function after this will print the array
with the shape of `(3,6)`. Finally, the last method is for assigning the shape value of
`(9,2)` to the `shape` attribute. This will change the shape to `(9,2)`.

The most important thing to remember is that while reshaping, the total size of the
new array should be unchanged:

```
import numpy as np
x2d = np.array((  (100,200,300),
                  (111,222,333),
                  (123,456,789),
                  (125,457,791),
                  (127,459,793),
                  (129,461,795) ))
print x2d
x2d.ravel()
print x2d
x2d.resize((3,6))
print x2d
x2d.reshape(6,3)
print x2d
x2d.shape = (9,2)
print x2d
```

If required, there are facilities to convert arrays into Python list data structures,
stored files, and strings. There are separate methods for each of these conversions
called `tolist`, `tofile`, and `tostring`.

Classes associated with arrays

There are a number of classes and subclasses associated with the `ndarray` class.
These classes are designed to support specific enhanced functionality. In the
following paragraphs, we will be discussing these classes and subclasses.

The matrix sub class

The matrix class is a Python subclass of ndarrays. A matrix can be created from other matrices or strings or any other object that can be converted into an ndarray. The matrix sub class has specially overwritten operators, such as * for matrix multiplication and ** for matrix power. Several functions are provided in the matrix class to perform various activities, such as sorting elements, calculation of the transpose, finding the sum of matrix elements, conversion of a matrix to a list, and other data structures and data types. Consider the following program, which defines two matrices with three rows and three columns each. At last, the program displays the output of matrix multiplication:

```
import numpy as np
a = np.matrix('1 2 3; 4 5 6; 7 8 9')
print a
b = np.matrix('4 5 6; 7 8 9; 10 11 12')
print b
print a*b
```

Masked array

NumPy has a module named numpy.ma for creating masked arrays. A masked array is a normal array that has some invalid, missing, or undesirable entries. It has two components: the original ndarrays and a mask. A mask is an array of Boolean values used to determine whether the array values are valid or not. A true value in the mask reflects that the corresponding value in the array is invalid. The masked, or invalid, entries will not be used in any further computation on masked arrays. The next program demonstrates the concept of masked arrays. Suppose, the original array x has the pulse rates of different persons and it has two invalid entries. To mask these invalid entries, the corresponding value is set to 1 (true) in mask. At the end, we compute the mean of the original and masked arrays. Without masking, the mean is 61.1, because of two negative values; after masking, the mean of the remaining eight values is 94.5:

```
import numpy as np
import numpy.ma as ma
x = np.array([72, 79, 85, 90, 150, -135, 120, -10, 60, 100])
mx = ma.masked_array(x, mask=[0, 0, 0, 0, 0, 1, 0, 1, 0, 0])
mx2 = ma.masked_array(x,mask=x<0)
x.mean()
mx.mean()
mx2.mean()
```

The structured/recor array

The NumPy ndarray can hold record type values. To create an array of the `record` type, we first need to create a data type of the record, and then we will use this data type as the type of the elements of the array. This record data type can be defined using the `dtype` data type's definition, and then we can use this `dtype` in the array definition. Consider the following program, which creates an array of records that have the minimum, maximum, and average temperatures of cities. The `dtype` function has two components: the names of the fields and their formats. The formats used in this example are 32-bit integers (`i4`), 32-bit float (`f4`) and string of 30 or less characters (`a30`):

```
import numpy as np
rectype= np.dtype({'names':['mintemp', 'maxtemp', 'avgtemp', 'city'],
'formats':['i4','i4', 'f4', 'a30']})

a = np.array([(10, 44, 25.2, 'Indore'),(10, 42, 25.2, 'Mumbai'), (2,
48, 30, 'Delhi')],dtype=rectype)

print a[0]
print a['mintemp']
print a['maxtemp']
print a['avgtemp']
print a['city']
```

The universal function object

A **universal function (unfunc)** is a function that operates on ndarrays on an element-by-element basis. It also supports broadcasting, type casting, and a number of other important features. Broadcasting in NumPy is the process of operating on arrays of different shapes. Specially during arithmetic operations, the array with a smaller shape will be broadcast across the larger array to make their shape compatible. Universal functions are instances of the `ufunc` class of NumPy.

Attributes

There are several attributes that each universal function possesses, although the user cannot set the values of these attributes. The following are the attributes of a universal function. There are some informational attributes that universal functions possess:

- `__doc__`: This contains the doc string of the `ufunc` function. Its first part is dynamically generated on the basis of the name, number of inputs, and number of outputs. Its second part is defined at the time of function creation, and it is stored with the function.

- `__name__`: This is the name of the ufunc.
- `ufunc.nin`: This represents the total number of inputs.
- `ufunc.nout`: This represents the total number of outputs.
- `ufunc.nargs`: This represents the total number of arguments, including inputs and outputs.
- `ufunc.ntypes`: This represents the total number of different types for which this function is defined.
- `ufunc.types`: This returns a list that has `ntypes` elements that have the types for which this function is defined.
- `ufunc.identity`: The identity value of this function.

Methods

All `ufuncs` have five methods, as given in the following list. The first four methods are relevant only to a ufunc that takes two input arguments and returns one output argument. These methods will raise a `ValueError` exception when they are attempted to call on other ufuncs. The fifth method allows the user to perform in-place operations using indexing. The following methods are available with each of the NumPy universal functions:

- `ufunc.reduce`: This reduces the array's dimension by one by applying ufunc along one axis.
- `ufunc.accumulate`: This accumulates the result of applying the operator to all elements.
- `ufunc.reduceat`: This performs reduce with the specified slices over a single axis.
- `ufunc.outer(A, B)`: This applies the ufunc operator to all *(a, b)* pairs for *a* in A and *b* in B.
- `ufunc.at`: This performs an unbuffered in-place operation on an operand for the specified elements.

Various available ufunc

There are a number of ufuncs (at present, more than 60) supported by NumPy. These functions cover a wide variety of operations, including simple mathematical operations (such as add, subtract, mod, and absolute), square, log, exponential, trigonometric, bitwise, comparison, and floating-point functions. Generally, it is better to use these functions instead of applying looping, as they are more efficient than looping.

The following program demonstrates the use of some of these ufuncs:

```
import numpy as np
x1 = np.array([72, 79, 85, 90, 150, -135, 120, -10, 60, 100])
x2 = np.array([72, 79, 85, 90, 150, -135, 120, -10, 60, 100])
x_angle = np.array([30, 60, 90, 120, 150, 180])
x_sqr = np.array([9, 16, 25, 225, 400, 625])
x_bit = np.array([2, 4, 8, 16, 32, 64])
np.greater_equal(x1,x2)
np.mod(x1,x2)
np.exp(x1)
np.reciprocal(x1)
np.negative(x1)
np.isreal(x1)
np.isnan(np.log10(x1))
np.sqrt(np.square(x_sqr))
np.sin(x_angle*np.pi/180)
np.tan(x_angle*np.pi/180)
np.right_shift(x_bit,1)
np.left_shift (x_bit,1)
```

In Python, if there is a value that cannot be represented as a number, then that value is called `nan`. For example, if we operate the `log10` ufunc on the `x1` array in the preceding program, then as output, there are to `nan`. There is a ufunc called `isnan` that verifies that the input argument is `nan`. Trigonometric functions requires arguments as an angular value in degrees. Normal decimal values are radians value that can be converted to degree by multiplying by 180/NumPy.pi. The bitwise left shift by 1 performs fast multiplication by a value of 2 to the argument. Similarly, the bitwise right shift by 1 performs fast division by a value of 2 to the argument. Generally, these ufuncs operate on arrays, and if there is any non-array argument, then that argument is broadcast as an array. Then, perform the element-by-element operation. This is the case in the last four lines of the preceding program.

The NumPy mathematical modules

NumPy has added modules for specific functionalities, for example, linear algebra, discrete Fourier transforms, random sampling, and the matrix algebra library. These functionalities are bundled in separate modules, as follows:

* `numpy.linalg`: This module supports various functionalities of linear algebra, such as inner, outer, and dot products of arrays and vectors; norms of vector and matrix; solutions of linear matrix equations; and methods of matrix inversion.

- numpy.fft: Discrete Fourier transforms have a wide range of applications in digital signal processing. This module has functions for computing various types of discrete Fourier transforms, including one-dimensional, two-dimensional, multidimensional, inverse, and Hermitian Fourier transforms.

- numpy.matlib: This module contains functions that, by default, return a matrix object instead of ndarrays. These functions include empty, zeros, ones, eye, rapmat, rand, randn, bmat, mat, and matrix.

- numpy.random: This module contains functions for performing random sampling from the specific population. There are functions for generating simple random data from the given population or range. It also supports the generation of random permutations. Furthermore, it has a range of functions that support various statistical-distribution-based generations of random sampling data.

The next program demonstrates the use of some functions from the linalg module. It computes the norm, inverse, determinant, eigenvalues, and right eigenvectors of a square matrix. It also demonstrates the linear equation solver by solving a system of two equations, *2x+3y=4* and *3x+4y=5*, which is done by representing them as an array. The allclose function in the last line compares the two arrays passed to it and returns true if they are equal element-wise within a tolerance limit. The eig method computes the eigenvalues and eigenvectors of a square array. The returned values are as follows: w is the eigenvalue and v is the eigenvector, where the v[:,i] column is the eigenvector of w[i]:

```
import numpy as np
from numpy import linalg as LA
arr2d = np.array((  (100,200,300),
          (111,222,333),
          (129,461,795) ))
eig_val, eig_vec = LA.eig(arr2d)
LA.norm(arr2d)
LA.det(arr2d)
LA.inv(arr2d)
arr1 = np.array([[2,3], [3,4]])
arr2 = np.array([4,5])
results = np.linalg.solve(arr1, arr2)
print results
np.allclose(np.dot(arr1, results), arr2)
```

Random sampling is an important aspect of scientific and commercial computing. The following program demonstrates some functions from each of the categories of functions supported by the `numpy` random sampling module. Besides size and population, some distributions require some statistical values, such as mean, mode, and standard deviation. The `permutation` function randomly permutes a sequence or returns a permuted range, whereas the `randint` function returns randomly selected elements from the range given by the first two arguments; the total number of elements will be given as the third argument. The remaining methods return samples from specific distributions, such as chi-square, Pareto, standard normal, and log normal:

```
import numpy as np
np.random.permutation(10)
np.random.randint(20,50, size=10)
np.random.random_sample(10)
np.random.chisquare(5,10) # degree of freedom, size
alpha, location_param = 4., 2.
s = np.random.pareto(alpha, 10) + location_param

s = np.random.standard_normal(20)

mean, std_deviation = 4., 2.
s = np.random.lognormal(mean, std_deviation, 10)
```

Introduction to SciPy

SciPy contains a number of submodules dedicated to the common functionality required by various scientific computing applications. The SciPy community recommends that scientists first check whether a required functionality has already been implemented before actually implementing it in SciPy. As almost all of the essential functionality of scientific computing has already been implemented, this checking will save the efforts that the scientists would have applied in reinventing the wheel. Moreover, the SciPy modules have been optimized and well-tested for bugs and possible errors. Hence, using them will be beneficial in terms of better performance.

Mathematical functions in SciPy

SciPy is written on top of NumPy to extends its functionality to perform advanced mathematical functionality. Basic mathematical functions available in NumPy are not redesigned to perform these functionalities. We need to use NumPy functions, as we will see, in the programs in the subsequent discussion in this chapter.

Advanced modules/packages

The functionality of SciPy is divided into a number of separate task-specific modules. Let's discuss these modules one by one. For brevity, we will not cover all the functions of any module. Instead, we will demonstrate some examples from each module of SciPy.

Integration

The `scipy.integrate` sub package has functions for several integration methods, including the integrator for ordinary differential equations. There are several methods for integrating functions when the function object is given. It has methods for integrating functions when fixed samples are given.

Here are the integrating functions for given function objects:

- `quad`: General-purpose integration
- `dblquad`: General-purpose double integration
- `tplquad`: General-purpose triple integration
- `nquad`: General-purpose n-dimensional integration
- `fixed_quad`: Integrate *func(x)* using a Gaussian quadrature of order *n*
- `quadrature`: Integrate within a given tolerance using a Gaussian quadrature
- `romberg`: Integrate func using Romberg integration

These are integrating functions for given fixed samples:

- `cumtrapz`: Use the trapezoidal rule to cumulatively compute the integral
- `simps`: Use Simpson's rule to compute the integral from the samples
- `romb`: Use Romberg integration to compute the integral from *(2**k + 1)* evenly spaced samples

The integrators for ordinary differential equation systems are as follows:

- `odeint`: General integration of ordinary differential equations
- `ode`: Integrate ODE using VODE and ZVODE routines
- `complex_ode`: Convert a complex-valued ODE to a real-valued and integrate

Let's discuss programs for selected methods from the preceding list. The `quad` function performs general integration of a function of one variable between two points within the range of plus or minus infinity. In the following program, we use the function to calculate the integral of the Bessel function of first kind for an interval of *(0,20)*. The Bessel function of first kind is defined in the `special.jv` method. The last line of the following program computes the Gaussian integral using the `quad` function:

```
import numpy as np
from scipy import special
from scipy import integrate

result = integrate.quad(lambda x: special.jv(4,x), 0, 20)
print result
print "Gaussian integral", np.sqrt(np.pi),quad(lambda x: np.exp(-
x**2),-np.inf, np.inf)
```

If the function to be integrated requires additional parameters, such as multiplication or power factors for variables, then these parameters can be passed as arguments. This is demonstrated in the following program, by passing a, b, and c as arguments to the `quad` function. Sometimes, it is possible that the integral is divergent or converges very slowly:

```
import numpy as np
from scipy.integrate import quad

def integrand(x, a, b, c):
    return a*x*x+b*x+c

a = 3
b = 4
c = 1
result = quad(integrand, 0,np.inf,  args=(a,b,c))
print result
```

Double and triple integration can be performed using the `dblquad` and `tplquad` functions, respectively. The next program demonstrates the use of the `dblquad` function. The t and x arguments vary from 0 to infinity (`inf`). The code after the comment performs the Gaussian quadrature over a fixed interval:

```
import numpy as np
from scipy.integrate import quad, dblquad, fixed_quad

def integrand1 (t, x, n):
    return np.exp(-x*t) / t**n
```

```
n = 4
result = dblquad(lambda t, x: integrand1(t, x, n), 0, np.inf, lambda
x: 0, lambda x: np.inf)
# the following code is performing Gaussian quadrature over a fixed
interval
from scipy.integrate import fixed_quad, quadrature

def integrand(x, a, b):
  return a * x + b
a = 2
b = 1
fixed_result = fixed_quad(integrand, 0, 1, args=(a,b))
result  = quadrature(integrand, 0, 1, args=(a,b))
```

For integrating a function with an arbitrarily spaced sample, we have the `simps` function. Simpson's rule approximates the function between three adjacent points as a parabola. The following program uses the `simps` function:

```
import numpy as np
from scipy.integrate import simps
def func1(a,x):
  return a*x**2+2

def func2(b,x):
  return b*x**3+4

x = np.array([1, 2, 4, 5, 6])
y1 = func1(2,x)
Intgrl1 = simps(y1, x)

print(Intgrl1)

y2 = func2(3,x)
Intgrl2 = simps (y2,x)
print (Intgrl2)
```

Here is a program that demonstrates the integration of ordinary differential equations using the `odeint` function:

```
import matplotlib.pyplot as plt
from numpy import linspace, array
def derivative(x,time):
  a = -2.0
  b = -0.1
  return array([  x[1], a*x[0]+b*x[1] ])
```

```
time = linspace (1.0,15.0,1000)
xinitialize = array ([1.05,10.2])
x = odeint(derivative,xinitialize,time)
plt.figure()
plt.plot(time,x[:,0])
plt.xlabel('t')
plt.ylabel('x')
plt.show()
```

Signal processing (scipy.signal)

The signal processing toolbox contains a number of filtering functions, filter-designing functions, and functions for several B-spline interpolation algorithms for one- and two-dimensional data. This toolbox has several functions for performing the following operations:

- Convolution
- B-splines
- Filtering
- Filter design
- Matlab-style IIR filter design
- Continuous-time linear systems
- Discrete-time linear systems
- LTI representations
- Waveforms
- Window functions
- Wavelets
- Peak finding
- Spectral analysis

Let's discuss some example programs to understand the functionalities of the signal processing toolbox.

The detrend function is a filtering function that removes constant or linear trends along the axis from the data so that we can see the effect of the higher order, as demonstrated in the following program:

```
import numpy as np
import matplotlib as mpl
import matplotlib.pyplot as plt
```

```
from scipy import signal
t = np.linspace(0, 5, 100)
x = t + np.random.normal(size=100)
plt.plot(t, x, linewidth=3)
plt.show()
plt.plot(t, signal.detrend(x), linewidth=3)
plt.show()
```

The following program uses spline filtering to compute an edge image of Lena's face taken as an array using the `misc.lena` command. This functionality is achieved by using two functions. First, the `cspline2d` command is used to apply a separable two-dimensional FIR filter with mirror-symmetric boundary conditions to the spline coefficients. This function is faster than the second function, `convolve2d`, which convolves arbitrary two-dimensional filters and permits you to choose mirror-symmetric boundary conditions:

```
import numpy as np
from scipy import signal, misc
import matplotlib.pyplot as plt
img = misc.lena()

splineresult = signal.cspline2d(img, 2.0)
arr1 = np.array([[-1,0,1], [-2,0,2], [-1,0,1]], dtype=np.float32)
derivative = signal.convolve2d(splineresult,arr1,boundary='symm'
,mode='same')
plt.figure()
plt.imshow(derivative)
plt.title('Image filtered by spline edge filter')
plt.gray()
plt.show()
```

Fourier transforms (scipy.fftpack)

The discrete Fourier transform and discrete inverse Fourier transform of a real or complex sequence can be calculated using `fft` and `ifft` (fast Fourier transform), respectively, as demonstrated in this program:

```
import numpy as np
from scipy.fftpack import fft, ifft
x = np.random.random_sample(5)
y = fft(x)
print y
yinv = ifft(y)
print yinv
```

The following program plots the FFT of the sum of three sines:

```
import numpy as np
import matplotlib as mpl
import matplotlib.pyplot as plt
from scipy.fftpack import fft
x = np.linspace(0.0, 1, 500)
y = np.sin(100*np.pi*x) + 0.5*np.sin(150*np.pi*x) + 0.75*np.
sin(200*np.pi*x)
yf = fft(y)
xf = np.linspace(0.0, 0.1, 250)
import matplotlib.pyplot as plt
plt.plot(xf, 2.0/500 * np.abs(yf[0:500/2]))
plt.grid()
plt.show()
```

Spatial data structures and algorithms (scipy.spatial)

Spatial analysis is a set of techniques and algorithms used for analysis of spatial data. The data objects or elements that are related to the geographical space or horizon can be called spatial data. This data consists of points, lines, polygons, and other geometrical and geographical primitives that can be mapped by locations and used to track and locate various devices. It may be scalar or vector data that provides specific information about a geographical or spatial location. Spatial data is used and processed by a number of applications in different areas, such as geography, geographical information systems/retrieval, location-based services, web- and desktop-based spatial applications, spatial mining, and others.

A **k-dimensional tree** (**k-d tree**) is a space partitioning data structure. It organizes points in a k-dimensional space. In mathematics, space partitioning is the process of dividing a space into multiple disjoint spaces. It divides the space into non-overlapping regions, where each point in the space may belong to exactly one region.

SciPy has a spatial module that supports various desired functionalities for spatial computing. The user can compute Delaunay triangulations, Voronoi diagrams, and convex hulls in N dimensions. It also has plotting helpers for plotting these computations in two dimensions. Moreover, to perform quick nearest neighbor lookups, it also supports the *KDTree functionality*, and has the facility for computation of the distant matrix from a collection of raw observation vectors.

Let's discuss some sample programs that show these functions. The following program performs Delaunay triangulation and then plots the results of the computation using the `pyplot`:

```
import numpy as np
import matplotlib.pyplot as plt
from scipy.spatial import Delaunay
arr_pt = np.array([[0, 0], [0, 1.1], [1, 0], [1, 1]])
arr1 = np.array([0., 0., 1., 1.])
arr2 = np.array([0., 1.1, 0., 1.])

triangle_result = Delaunay(arr_pt)
plt.triplot(arr1, arr2, triangle_result.simplices.copy())
plt.show()
plt.plot(arr1, arr2, 'ro')
plt.show()
```

The smallest convex object that contains all the points of a given point set is called a convex hull, and it can be computed using the `convexHull` function. The next program demonstrates the use of this function and then plots the results of computing using the `convexHull` function:

```
import numpy as np
from scipy.spatial import ConvexHull
import matplotlib.pyplot as plt
randpoints = np.random.rand(25, 2)
hull = ConvexHull(randpoints)
#following line will draw points
plt.plot(randpoints[:,0], randpoints[:,1], 'x')
#this loop will draw the line segment
for simplex in hull.simplices:
    plt.plot(randpoints[simplex,0], randpoints[simplex,1], 'k')

plt.show()
```

We can use KDTree to find out which point from the set of points is closest to the selected point. This program demonstrates the use of the k-d tree:

```
from scipy import spatial
x_val, y_val = np.mgrid[1:5, 3:9]
tree_create = spatial.KDTree(zip(x_val.ravel(), y_val.ravel()))
tree_create.data
points_for_query = np.array([[0, 0], [2.1, 2.9]])
tree_create.query(points_for_query)
```

The following program displays the closest distance and the indices:

```
import numpy as np
import matplotlib.pyplot as plt
from scipy.spatial import KDTree
vertx = np.array([[1, 1], [1, 2], [1, 3], [2, 1], [2, 2], [2, 3], [3,
1], [3, 2], [3, 3]])
tree_create = KDTree(vertx)
tree_create.query([1.1, 1.1])
x_vals = np.linspace(0.5, 3.5, 31)
y_vals = np.linspace(0.5, 3.5, 33)
xgrid, ygrid = np.meshgrid(x, y)
xy = np.c_[xgrid.ravel(), ygrid.ravel()]
plt.pcolor(x_vals, y_vals, tree.query(xy)[1].reshape(33, 31))
plt.plot(points[:,0], points[:,1], 'ko')
plt.show()
```

Optimization (scipy.optimize)

Optimization is the process to find the best solution for an objective function of one or more variables and possibly in presence of some predefined constraints on these variables, and some possibly in the presence of some predefined constraints on these variables. The objective function may be considered as a cost/energy function to be minimized or a profit or utility function to be maximized. There are a few important concepts associated with optimization problems, such as the dimensionality of the optimization problem and the type of optimization. Before solving an optimization problem, it is better to understand these concepts first and then start working on the solution. By dimensionality of the problem, we mean the number of scalar variables on which the search for the optimized value is to be performed. The number of variables can be one or more. This number of variables also affects the scalability of the solution. The more the number of scalar variables, the slower the problem. Moreover, the type of optimization also has an impact on the designing of the solution.

Another important consideration is whether the problem is a constrained problem or not. By constrained problem, we mean that the solution must fulfill some predefined constraints on the variables under study. For example, we can write a general-constraint minimization optimization problem as follows:

```
Minimize                         f(x)
Subjected to constraints    gi (x)= ai     for i= 1 … … n
                            Hj (x)>= bj      for j= 1 … … m
```

These constraints are required to be satisfied by the solution. The solution of the problem depends on the relationship between the objective function, the constraints, and the variables. Moreover, the size of the model also affects the solution. The size of the model is measured by the number of variables and the number of constraints in it. Generally, there is an upper limit of the size of model, which is imposed by most optimization solver software applications. This limit has to be introduced due to the higher memory requirements, the processing demands of the problem, and its numerical stability. There might be a possibility that we don't find a solution at all, or the getting the solution may be very time-consuming, which gives the impression that the solution isn't converging.

Furthermore, an optimization problem may be a convex or non-convex problem. A convex problem is comparatively simpler to solve, as it has one global minimum/maximum and no local minimum/maximum.

Let's discuss the concept of convexity in detail. Convex optimization is the process of minimizing a `convex` function over a convex set. The `convex` function (it holds real values) defined for an interval is called a `convex` function if a line segment between any two points on the graph lies on or above the graph. Two of the popular `convex` functions are the exponential function ($f(x)=ex$) and quadratic function ($f(x)=x2$). Some examples of `convex` and non-convex functions are shown in this figure:

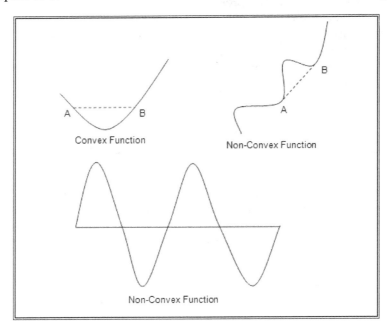

Now, the convex set is such a region in which if we join two points inside it by a line segment, then all the points on that line segment also lie inside the region. The following figure depicts the convex and non-convex sets:

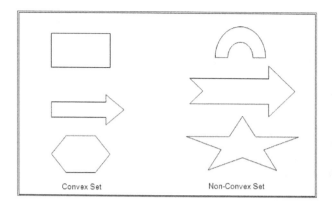

Convex Set   Non-Convex Set

The scipy.optimize package provides functions for most of the useful algorithms for scalar and multidimensional function minimization, curve fitting, and root finding. Let's discuss how to use these functions.

The next program demonstrates the use of the **Broyden-Fletcher-Goldfarb-Shanno (BFGS)** algorithm. This algorithm uses the gradient of the objective function to quickly converge to the solution. The program first defines a function called rosen_derivative to compute the gradient of the rosenbrock function:

```
import numpy as np
from scipy.optimize import minimize
def rosenbrock(x):
   return sum(100.0*(x[1:]-x[:-1]**2.0)**2.0 + (1-x[:-1])**2.0)

x0 = np.array([1.3, 0.7, 0.8, 1.9, 1.2])

def rosen_derivative(x):
  x1 = x[1:-1]
  x1_m1 = x[:-2]
  x1_p1 = x[2:]
  derivative = np.zeros_like(x)
  derivative[1:-1] = 200*(x1-x1_m1**2) - 400*(x1_p1 - x1**2)*x1 -
2*(1-x1)
  derivative[0] = -400*x[0]*(x[1]-x[0]**2) - 2*(1-x[0])
  derivative[-1] = 200*(x[-1]-x[-2]**2)
  return derivative

res = minimize(rosenbrock, x0, method='BFGS', jac=rosen_derivative,
options={'disp': True})
```

The following program first computes the Hessian of the Rosenbrock function, and then minimizes the function using the Newton conjugate gradient method:

```python
import numpy as np
from scipy.optimize import minimize

def rosenbrock(x):
  return sum(100.0*(x[1:]-x[:-1]**2.0)**2.0 + (1-x[:-1])**2.0)

x0 = np.array([1.3, 0.7, 0.8, 1.9, 1.2])

def rosen_derivative(x):
  x1 = x[1:-1]
  x1_m1 = x[:-2]
  x1_p1 = x[2:]
  derivative = np.zeros_like(x)
  derivative[1:-1] = 200*(x1-x1_m1**2) - 400*(x1_p1 - x1**2)*x1 -
2*(1-x1)
  derivative[0] = -400*x[0]*(x[1]-x[0]**2) - 2*(1-x[0])
  derivative[-1] = 200*(x[-1]-x[-2]**2)
  return derivative

def rosen_hessian(x):
  x_val = np.asarray(x)
  hess = np.diag(-400*x_val[:-1],1) - np.diag(400*x_val[:-1],-1)
  diagonal = np.zeros_like(x_val)
  diagonal[0] = 1200*x_val[0]**2-400*x_val[1]+2
  diagonal[-1] = 200
  diagonal[1:-1] = 202 + 1200*x_val[1:-1]**2 - 400*x_val[2:]
  hess = hess + np.diag(diagonal)
  return hess

result = minimize(rosenbrock, x0, method='Newton-CG', jac=rosen_
derivative, hess=rosen_hessian, options={'xtol': 1e-8, 'disp': True})
print result.x
```

The `minimize` function also has an interface to a number of constrained minimization algorithms. The following program uses the **Sequential Least Square Programming optimization (SLSQP)** algorithm. The function to be minimized is defined in `func`, its derivative is defined in `func_deriv`, and the constraints are defined in the `cons` variable:

```python
import numpy as np
from scipy.optimize import minimize
def func(x, sign=1.0):
```

```
    return sign*(2*x[0]*x[1] + 2*x[0] - x[0]**2 - 2*x[1]**2)

def func_deriv(x, sign=1.0):
  dfdx0 = sign*(-2*x[0] + 2*x[1] + 2)
  dfdx1 = sign*(2*x[0] - 4*x[1])
  return np.array([ dfdx0, dfdx1 ])

cons = ({'type': 'eq',
  'fun': lambda x: np.array([x[0]**3 - x[1]]),
  'jac': lambda x: np.array([3.0*(x[0]**2.0), -1.0])},
  {'type': 'ineq',
  'fun': lambda x: np.array([x[1] - 1]),
  'jac': lambda x: np.array([0.0, 1.0])})

res = minimize(func, [-1.0,1.0], args=(-1.0,), jac=func_deriv,
method='SLSQP', options={'disp': True})
print(res.x)

res = minimize(func, [-1.0,1.0], args=(-1.0,), jac=func_
deriv,constraints=cons, method='SLSQP', options={'disp': True})
print(res.x)
```

The next program demonstrates the methods of finding the global minimum and local minimum. First, it defines the function and plots it. This function has a global minimum around -1.3 and a local minimum around 3.8. The BFGF algorithm is used to find the local minimum. The program uses a brute-force algorithm to find the global minimum. However, with the increase in grid size(the range/domain of values to be checked), the brute-force method becomes slow, so it is better to use the Brent method for scalar functions. The program also uses the fminbound function to find the local minimum between 0 and 10:

```
import numpy as np
import matplotlib.pyplot as plt
from scipy import optimize

def f(x):
  return x**2 + 10*np.sin(x)

x = np.arange(-10,10,0.1)
plt.plot(x, f(x))
plt.show()

optimize.fmin_bfgs(f, 0)
```

```
grid = (-10, 10, 0.1)
optimize.brute(f, (grid,))
optimize.brent(f)
optimize.fminbound(f, 0, 10)
```

The following program demonstrates the use of constrained optimization:

```
import numpy as np
from scipy import optimize
def f(x):
    return np.sqrt((x[0] - 2)**2 + (x[1] - 3)**2)

def constraint(x):
    return np.atleast_1d(2.5 - np.sum(np.abs(x)))

optimize.fmin_slsqp(f, np.array([0, 2]), ieqcons=[constraint, ])
optimize.fmin_cobyla(f, np.array([3, 4]), cons=constraint)
```

There are several methods for finding the roots of a polynomial, the next three programs use the bisection method, Newton-Raphson method, and root function. The bisection method is used in this program to find the roots of the polynomial defined in `polynomial_func`:

```
import scipy.optimize as optimize
import numpy as np

def polynomial_func(x):
    return np.cos(x)**3 + 4 - 2*x

print(optimize.bisect(polynomial_func, 1, 5))
```

The Newton-Raphson method is used to find the roots of a polynomial in the following program:

```
import scipy.optimize
from scipy import optimize

def polynomial_func(xvalue):
  yvalue = xvalue + 2*scipy.cos(xvalue)
        return yvalue

scipy.optimize.newton(polynomial_func, 1)
```

In mathematics, the Lagrange multipliers method is used to find the local minima and local maxima of a function, subject to equality constraints. This program computes the Lagrange multipliers using the `fsolve` method:

```python
import numpy as np
from scipy.optimize import fsolve
def func_orig(data):
    xval = data[0]
    yval = data[1]
    Multiplier = data[2]
    return xval + yval + Multiplier * (xval**2 + yval**2 - 1)

def deriv_func_orig(data):
    dLambda = np.zeros(len(data))
    step_size = 1e-3 # this is the step size used in the finite
difference.
    for i in range(len(data)):
        ddata = np.zeros(len(data))
        ddata[i] = step_size
        dLambda[i] = (func_orig(data+ddata)-func_orig(data-ddata))/
(2*step_size);
        return dLambda


data1 = fsolve(deriv_func_orig, [1, 1, 0])
print data1, func_orig(data1)

data2 = fsolve(deriv_func_orig, [-1, -1, 0])
print data2, func_orig(data2)
```

Interpolation (scipy.interpolate)

Interpolation is a method of finding new data points within the range of a discrete set of well-known data points. The `interpolate` subpackage has interpolators for computation using various interpolation methods. It supports interpolation using `spline` functions, `univariate` and `multivariate` one-dimensional and multidimensional interpolation, Lagrange and Taylor polynomial interpolators. It also has wrapper classes for the `FITPACK` and `DFITPACK` functions. Let's discuss some programs that demonstrate the use of some of these methods.

This program demonstrates one-dimensional interpolation using linear and cubic interpolation and plots them for comparison:

```python
import numpy as np
from scipy.interpolate import interp1d
x_val = np.linspace(0, 20, 10)
y_val = np.cos(-x**2/8.0)
f = interp1d(x_val, y_val)
f2 = interp1d(x_val, y_val, kind='cubic')
xnew = np.linspace(0, 20, 25)
import matplotlib.pyplot as plt
plt.plot(x,y,'o',xnew,f(xnew),'-', xnew, f2(xnew),'--')
plt.legend(['data', 'linear', 'cubic'], loc='best')
plt.show()
```

The following program demonstrates the use of the `griddata` function for multivariate data interpolation over 150 points. This number of points can be changed to any suitable value. The program uses `pyplot` to create four subplots in single plot:

```python
import numpy as np
import matplotlib.pyplot as plt
from scipy.interpolate import griddata

def func_user(x, y):
    return x*(1-x)*np.cos(4*np.pi*x) * np.sin(4*np.pi*y**2)**2

x, y = np.mgrid[0:1:100j, 0:1:200j]

points = np.random.rand(150, 2)
values = func_user(points[:,0], points[:,1])
grid_z0 = griddata(points, values, (x, y), method='nearest')
grid_z1 = griddata(points, values, (x, y), method='linear')
grid_z2 = griddata(points, values, (x, y), method='cubic')

f, axarr = plt.subplots(2, 2)
axarr[0, 0].imshow(func(x, y).T, extent=(0,1,0,1), origin='lower')
axarr[0, 0].plot(points[:,0], points[:,1], 'k', ms=1)
axarr[0, 0].set_title('Original')

axarr[0, 1].imshow(grid_z0.T, extent=(0,1,0,1), origin='lower')
axarr[0, 1].set_title('Nearest')

axarr[1, 0].imshow(grid_z1.T, extent=(0,1,0,1), origin='lower')
axarr[1, 0].set_title('Linear')
```

```
axarr[1, 1].imshow(grid_z2.T, extent=(0,1,0,1), origin='lower')
axarr[1, 1].set_title('Cubic')

plt.show()
```

Linear algebra (scipy.linalg)

The `scipy` linear algebra methods expect an argument as an object that can be converted to a two-dimensional array. The methods also return a two-dimensional array. The `scipy.linalg` function has advanced features in comparison to `numpy.linalg`.

The following program computes the inverse of matrix represented as a two-dimensional array. It also uses `T` (the shortcut for transpose) and performs multiplication over the array:

```
import numpy as np
from scipy import linalg
A = np.array([[2,3],[4,5]])
linalg.inv(A)
B = np.array([[3,8]])
A*B
A.dot(B.T)
```

This small program computes the inverse of a matrix and its determinant:

```
import numpy as np
from scipy import linalg
A = np.array([[2,3],[4,5]])
linalg.inv(A)
linalg.det(A)
```

The next program demonstrates the solving of linear equations using the matrix inverse and its fast implementation using the solver:

```
import numpy as np
from scipy import linalg
A = np.array([[2,3],[4,5]])
B = np.array([[5],[6]])
linalg.inv(A).dot(B)
np.linalg.solve(A,B)
```

The following program seeks a set of linear scaling coefficients and fits that data using a model. This program uses linalg.lstsq to solve the data fitting problem. The lstsq method is used to find the least square solutions of linear matrix equations. This method is a tool for finding the best fit line for the given data points. It uses linear algebra and simple calculus:

```
import numpy as np
from scipy import linalg
import matplotlib.pyplot as plt
coeff_1, coeff_2 = 5.0, 2.0
i = np.r_[1:11]  # or we can use np.arang(1, 11)
x = 0.1*i
y = coeff_1*np.exp(-x) + coeff_2*x
z = y + 0.05 * np.max(y) * np.random.randn(len(y))

A = np.c_[np.exp(-x)[:, np.newaxis], x[:, np.newaxis]]
coeff, resid, rank, sigma = linalg.lstsq(A, zi)

x2 = np.r_[0.1:1.0:100j]
y2 = coeff[0]*np.exp(-x2) + coeff[1]*x2

plt.plot(x,z,'x',x2,y2)
plt.axis([0,1,3.0,5.5])
plt.title('Data fitting with linalg.lstsq')
plt.show()
```

The following program demonstrates a method for singular value decomposition and the linag.svd and linag.diagsvd functions:

```
import numpy as np
from scipy import linalg
A = np.array([[5,4,2],[4,8,7]])
row = 2
col = 3
U,s,Vh = linalg.svd(A)
Sig = linalg.diagsvd(s,row,col)
U, Vh = U, Vh
print U
print Sig
print Vh
```

Sparse eigenvalue problems with ARPACK

This program computes the standard eigenvalue decomposition and the corresponding eigenvectors:

```
import numpy as np
from scipy.linalg import eigh
from scipy.sparse.linalg import eigsh
#following line is suppressing the values after decimal
np.set_printoptions(suppress=True)

np.random.seed(0)
random_matrix = np.random.random((75,75)) - 0.5
random_matrix = np.dot(random_matrix, random_matrix.T)
#compute eigenvalues decomposition
eigenvalues_all, eigenvectors_all = eigh(random_matrix)

eigenvalues_large, eigenvectors_large = eigsh(random_matrix, 3,
which='LM')
print eigenvalues_all[-3:]
print eigenvalues_large
print np.dot(eigenvectors_large.T, eigenvectors_all[:,-3:])
```

If we try for eigenvalues with the smallest values using `eigenvalues_small,` `eigenvectors_small = eigsh(random_matrix, 3, which='SM')`, in this case, the system returns with an error of no `convergence`. There are a few options for solving this problem. The first solution is to increase the tolerance limit by passing `tol=1E-2` to the `eigsh` function like this: `eigenvalues_small, eigenvectors_small = eigsh(random_matrix, 3, which='SM', tol=1E-2)`. This will solve the problem but lead to loss of precision.

Another solution is to increase maximum number of iterations to 5,000 by passing `maxiter=5000` to the `eigsh` function like this: `eigenvalues_small, eigenvectors_small = eigsh(random_matrix, 3, which='SM', maxiter=5000)`. However, more iterations will take longer, and there is a better way of solving this quickly with the desired precision. Use the shift-inter mode using the `sigma=0` or 2 and `which='LM'` arguments, as follows: `eigenvalues_small, eigenvectors_small = eigsh(random_matrix, 3, sigma=0, which='LM')`.

Statistics (scipy.stats)

There are a number of statistical functions that are designed to work with arrays, and their special versions are designed to work on masked arrays. The programs in subsequent paragraphs demonstrate the use of some of the available functions for continuous and discrete probability distributions.

The following program uses the discrete binomial random variable and plots its probability mass function. Here is the probability mass function of a binomial discrete distribution:

```
binom.pmf(k) = choose(n, k) * p**k * (1-p)**(n-k)
```

In the preceding code, where k is in (0,1,...,n) , n and p are shape parameters:

```
import numpy as np
from scipy.stats import binom
import matplotlib.pyplot as plt


n, p = 5, 0.4
mean, variance, skewness, kurtosis = binom.stats(n, p, moments='mvsk')
x_var = np.arange(binom.ppf(0.01, n, p),binom.ppf(0.99, n, p))

plt.plot(x_var, binom.pmf(x_var, n, p), 'ro', ms=5, label='PMF of
binomial ')
plt.vlines(x_var, 0, binom.pmf(x_var, n, p), colors='r', lw=3,
alpha=0.5)
plt.show()
```

The next program demonstrates the use of the geometric discrete random variable and plots the probability mass function:

```
geom.pmf(k) = (1-p)**(k-1)*p
```

Here, *k >= 1* and *p* is the shape parameter:

```
import numpy as np
from scipy.stats import geom
import matplotlib.pyplot as plt


p = 0.5
mean, variance, skewness, kurtosis = geom.stats(p, moments='mvsk')
x_var = np.arange(geom.ppf(0.01, p),geom.ppf(0.99, p))
plt.plot(x_var, geom.pmf(x_var, p), 'go', ms=5, label='PMF of
geomatric')
plt.vlines(x_var, 0, geom.pmf(x_var, p), colors='g', lw=3, alpha=0.5)

plt.show()
```

The following program demonstrates the computation of a generalized Pareto continuous random variable and plots its probability density function:

```
genpareto.pdf(x, c) = (1 + c * x)**(-1 - 1/c)
Here, x >= 0 if c >=0 and 0 <= x <= -1/c if c < 0:
```

```
import numpy as np
from scipy.stats import genpareto
import matplotlib.pyplot as plt
c = 0.1
mean, variance, skewness, kurtosis  = genpareto.stats(c,
moments='mvsk')
x_val = np.linspace(genpareto.ppf(0.01, c),genpareto.ppf(0.99, c),
100)
plt.plot(x_val, genpareto.pdf(x_val, c),'b-', lw=3, alpha=0.6,
label='PDF of Generic Pareto')
plt.show()
plt.show()
```

The next program shows the computation of a generalized gamma continuous random variable and plots its probability density function:

```
gengamma.pdf(x, a, c) = abs(c) * x**(c*a-1) * exp(-x**c) / gamma(a)
```

This time, r $x > 0$, $a > 0$, and $c!= 0$. Here, a and c are shape parameters:

```
import numpy as np
from scipy.stats import gengamma
import matplotlib.pyplot as plt
a, c = 4.41623854294, 3.11930916792
mean, variance, skewness, kurtosis  = gengamma.stats(a, c,
moments='mvsk')
x_var = np.linspace(gengamma.ppf(0.01, a, c),gengamma.ppf(0.99, a, c),
100)
plt.plot(x_var, gengamma.pdf(x_var, a, c),'b-', lw=3, alpha=0.6,
label='PDF of generic Gamma')
plt.show()
```

The following program demonstrates the computation of a multivariate normal random variable and plots its probability density function. For simplicity, we are skipping the probability density function:

```
import numpy as np
import matplotlib.pyplot as plt
from scipy.stats import multivariate_normal
x_var = np.linspace(5, 25, 20, endpoint=False)
y_var = multivariate_normal.pdf(x_var, mean=10, cov=2.5)
plt.plot(x_var, y_var)
plt.show()
```

We can also freeze these statistical distributions to display the frozen probability distribution / mass function.

Multidimensional image processing (scipy.ndimage)

Generally, image processing and image analysis can be considered as performing operations on two-dimensional arrays of values. This package provides a number of image processing and image analysis functions to be applied on arrays. The following code works on the image of Lena. First, the program introduces some noises into the image, and then it uses some filters to clean the noise. It displays the noisy image and the filtered image using the Gaussian, median, and `signal.wiener` filters:

```python
import numpy as np
from scipy import signal
from scipy import misc
from scipy import ndimage
import matplotlib.pyplot as plt

lena = misc.lena()
noisy_lena = np.copy(lena).astype(np.float)
noisy_lena += lena.std()*0.5*np.random.standard_normal(lena.shape)
f, axarr = plt.subplots(2, 2)
axarr[0, 0].imshow(noisy_lena, cmap=plt.cm.gray)
axarr[0, 0].axis('off')
axarr[0, 0].set_title('Noissy Lena Image')
blurred_lena = ndimage.gaussian_filter(noisy_lena, sigma=3)
axarr[0, 1].imshow(blurred_lena, cmap=plt.cm.gray)
axarr[0, 1].axis('off')
axarr[0, 1].set_title('Blurred Lena')
median_lena = ndimage.median_filter(blurred_lena, size=5)
axarr[1, 0].imshow(median_lena, cmap=plt.cm.gray)
axarr[1, 0].axis('off')
axarr[1, 0].set_title('Median Filter Lena')
wiener_lena = signal.wiener(blurred_lena, (5,5))
axarr[1, 1].imshow(wiener_lena, cmap=plt.cm.gray)
axarr[1, 1].axis('off')
axarr[1, 1].set_title('Wiener Filter Lena')
plt.show()
```

The output of the preceding program is presented in this screenshot:

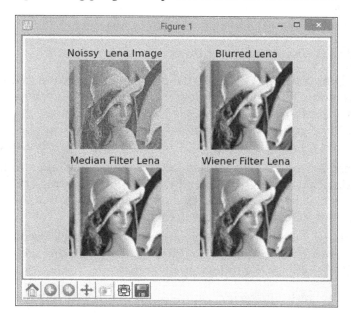

Clustering

Clustering is the process of putting a large set of objects into multiple groups. It uses some parameters in such a way that the objects in one group (known as a cluster) are more similar to each other than the objects in other groups or clusters.

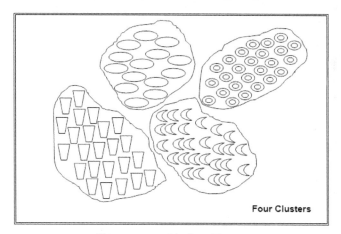

Objects grouped in four Clusters

The SciPy clustering package has two modules: **Vector Quantization (VQ)** and hierarchy. The VQ module supports k-means and vector quantization. The hierarchy module supports hierarchical and agglomerative clustering.

Let's get a brief idea about these algorithms:

- **Vector quantization**: VQ is a signal processing technique that enables its users to model the probability density functions by distribution of prototype vectors. It performs this modeling by dividing a large set of vectors or points into multiple groups having approximately the same number of points in their vicinity. Each of these groups has a representative centroid point.

- **k-means**: k-means is a vector quantization technique taken from signal processing that is widely used and popular for clustering analysis. It partitions *n* observations into *k* clusters in such a way that each observation belongs to the cluster with the nearest mean.

- **Hierarchical clustering**: This clustering technique seeks to build a hierarchy of clusters from the observations. Hierarchical clustering techniques generally belongs to the following two types:

  - **Divisible clustering**: This is a top-down approach to creating the hierarchy of clusters. It starts with one topmost cluster and performs splitting while moving downward.

  - **Agglomerative clustering**: This is a bottom-up approach. Each observation is a cluster, and this technique performs pairing of such clusters while moving up.

Generally, the results of hierarchical clustering are depicted in a dendrogram, that is, a tree diagram, as follows:

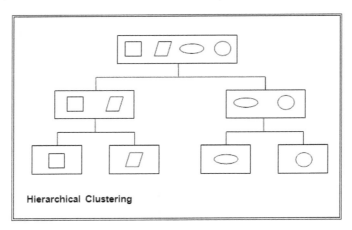

Hierarchical Clustering

The following program demonstrates an example of k-means clustering and vector quantization:

```
from scipy.cluster.vq import kmeans,vq
from numpy.random import rand
from numpy import vstack,array
from pylab import plot,show

data_set = vstack((rand(200,2) + array([.5,.5]),rand(200,2)))

# K-Means computation for 2 clusters
centroids_of_clusters,_ = kmeans(data_set,2)
index,_ = vq(data_set,centroids_of_clusters)


plot(data_set[index==0,0],data_set[index==0,1],'ob',
     data_set[index==1,0],data_set[index==1,1],'or')
plot(centroids_of_clusters[:,0],centroids_of_clusters[:,1],'sg',marke
rsize=8)

show()

# The same data for 3 clusters
centroids_of_clusters,_ = kmeans(data_set,3)
index,_ = vq(data_set,centroids_of_clusters)

plot(data_set[index==0,0],data_set[index==0,1],'ob',
     data_set[index==1,0],data_set[index==1,1],'or',
     data_set[index==2,0],data_set[index==2,1],'og') # third cluster
points
plot(centroids_of_clusters[:,0],centroids_of_clusters[:,1],'sm',marke
rsize=8)
show()
```

The hierarchical clustering module has a number of functions divided into many categories, such as functions for cutting hierarchical clustering into flat clustering, routines for agglomerating clustering, routines for visualization of clusters, data structures, and routines for representing hierarchies as tree structures, routines for computing statistics on hierarchies, predicate functions for checking the validity of linkage and inconstancy metrics, and so on. The following programs are used to draw a dendrogram of sample data using the linkage (agglomerative clustering) and dendrogram functions of the hierarchical module:

```
import numpy
from numpy.random import rand
from matplotlib.pyplot import show
```

```
from scipy.spatial.distance import pdist
import scipy.cluster.hierarchy as sch

x = rand(8,80)
x[0:4,:] *= 2

y = pdist(x)
z = sch.linkage(y)
sch.dendrogram(z)
show()
```

The output is depicted as follows:

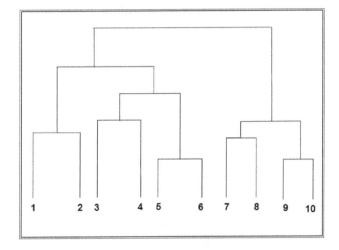

Curve fitting

The process of the construction of a mathematical function or curve that has the best fit for a series of data points is known as curve fitting. Generally, this curve fitting is subject to some constraints. The output of curve fitting can be used for data visualization to gain insights into a function when no data is available. Curve fitting can also be used to observe relationships among multiple variables. We can use curve fitting for different types of curves, such as lines, polynomials, conic sections, trigonometric functions, circles, ellipses, and others.

This program first creates some random data with noise. Then, it defines a function that represents the model (`line_func`) and performs curve fitting. Next, it determines the actual parameters, `a` and `b`. Finally, it also plots the errors:

```python
import numpy as np
import matplotlib.pyplot as plt
from scipy.optimize import curve_fit

xdata = np.random.uniform(0., 50., 80)
ydata = 3. * xdata + 2. + np.random.normal(0., 5., 80)
plt.plot(xdata, ydata, '.')

def line_func(x, a, b):
    return a * x + b

opt_param, cov_estimated = curve_fit(line_func, xdata, ydata)

errors = np.repeat(5., 80)
plt.errorbar(xdata, ydata, yerr=errors, fmt=None)

opt_param, cov_estimated = curve_fit(line_func, xdata, ydata,
sigma=errors)

print "a =", opt_param[0], "+/-", cov_estimated[0,0]**0.5
print "b =", opt_param[1], "+/-", cov_estimated[1,1]**0.5

plt.errorbar(xdata, ydata, yerr=errors, fmt=None)

xnew = np.linspace(0., 50., 80)
plt.plot(xnew, line_func(xnew, opt_param[0], opt_param[1]), 'r-')
plt.errorbar(xdata, ydata, yerr=errors, fmt=None)
plt.plot(xnew, line_func(xnew, *opt_param), 'r-')
plt.show()
```

The following program fits the curve for the *cos* trigonometric function:

```python
import numpy as np
from scipy import optimize
import pylab as pl

np.random.seed(0)

def func(x, omega, p):
    return np.cos(omega * x + p)
```

```
x = np.linspace(0, 10, 100)
y = f(x, 2.5, 3) + .1*np.random.normal(size=100)
params, params_cov = optimize.curve_fit(f, x, y)
t = np.linspace(0, 10, 500)
pl.figure(1)
pl.clf()
pl.plot(x, y, 'bx')
pl.plot(t, f(t, *params), 'r-')
pl.show()
```

File I/O (scipy.io)

SciPy provides support for performing read and write data to a variety of file formats using modules, classes, and functions:

- Matlab files
- ALD files
- Matrix market files
- Unformatted FORTRAN files
- WAV sound files
- ARFF files
- NetCDF files

This program performs reading and writing on NetCDF files:

```
from scipy.io import netcdf
# file creation
f = netcdf.netcdf_file('TestFile.nc', 'w')
f.history = 'Test netCDF File Creation'
f.createDimension('age', 12)
age = f.createVariable('age', 'i', ('age',))
age.units = 'Age of persons in Years'
age[:] = np.arange(12)
f.close()

#Now reading the file created
f = netcdf.netcdf_file('TestFile.nc', 'r')
print(f.history)
age = f.variables['age']
print(age.units)
print(age.shape)
print(age[-1])
f.close()
```

In a similar way, we can perform read/write operations on files of various other types. There is a separate submodule for loading the files created by the WEKA machine learning tools. WEKA stores files in the ARFF format, which is the standard format for WEKA. ARFF is a text file that may contain numerical, string, and data values. The following program reads and displays the data stored in the test.arff file. The content of the file is @relation foo; @attribute width numeric; @attribute height numeric; @attribute color {red,green,blue,yellow,black}; @data; 5.0,3.25,blue; 4.5,3.75,green; 3.0,4.00,red.

The program for reading and displaying the content is as follows:

```
from scipy.io import arff

file1 = open('test.arff')
data, meta = arff.loadarff(file1)

print data
print meta
```

Summary

In this chapter, we extensively discussed how to perform numerical computations using the NumPy and Scipy packages of Python. The concepts were presented along with example programs. The chapter started with a discussion on the fundamental objects of NumPy, and then we moved on to the advanced concepts of NumPy.

This chapter also discussed the functions and modules of SciPy. It covered the basic and special functions provided by SciPy, and then covered the special modules or sub packages. This was for showing advanced concepts, such as optimization, interpolation, Fourier transformation, signal processing, linear algebra, statistics, spatial algorithms, image processing, and file input and output.

In the next chapter, we will have an exhaustive discussion on symbolic computing, or CAS, using SymPy. Specifically, we will cover the core capabilities and extended functionalities for polynomials, calculus, equation solvers, discrete mathematics, geometry, and physics.

6
Applying Python for Symbolic Computing

SymPy includes functionality ranging from basic symbolic arithmetic to polynomials, calculus, solvers, discrete mathematics, geometry, statistics, and physics. It mainly works on three types of numbers, namely integer, real, and rational. Integers are whole digit numbers without a decimal point, while real numbers are numbers with decimal points. Rational numbers have two parts: the numerator and the denominator. To define rational numbers, we can use the Ration class, which requires two numbers. In this chapter, we will discuss the concepts of SymPy with the help of example programs.

We will cover the following topics in this chapter:

- A computerized algebra system using SymPy
- Core capabilities and advanced functionality
- Polynomials, calculus, and solving equations
- Discrete mathematics, matrices, geometry, plotting, physics, and statistics
- The printing functionality

Let's start a discussion on SymPy and its core capabilities, including basic arithmetic, expansion, simplification, substitution, pattern matching, and various functions (for example, exponential, logarithms, roots of equations, trigonometric functions, hyperbolic functions, and special functions).

Symbols, expressions, and basic arithmetic

In SymPy, we need to define symbols before using them in any expression. Defining a symbol is very simple. We just need to use the `symbol` function from the `Symbol` class to define a symbol, as used in the following program. We can use the `evalf()`/`n()` method to get the `float` numerical approximation of any object.

The following program uses three ways to create symbols. For creating only one symbol the name of method is symbol and for creating multiple symbols the method name is symbols. There are two ways of creating multiple symbols: one is by passing space-separated symbol names to the symbols method, and the other is by creating a sequence of symbols such as `m0, m1, m2, m3, m4` by passing `m0:5` to the symbols method. In this second option, `m0` is the first value of index and the number `5` after `:` denotes that a total of five such variables should be created.

Generally, division of two integers truncates the decimal part. To avoid this, the first line of the following program forces the performance of actual floating-point division on two integers. That's why the last line of the program will store `3.142857142857143` in `y`. If we ignore the first line in the following program. then the value of `y` will be `3`:

```
from __future__ import division
from sympy import *

x, y, z = symbols('x y z')
m0, m1, m2, m3, m4 = symbols('m0:5')
x1 = Symbol('x1')
x1 + 500
y=22/7
```

The following program uses the `evalf()` and `n()` methods to numerically approximate any SymPy object into a `float` value. The default accuracy is up to 15 decimal digits, and it can be changed to any desired accuracy by passing an argument to these methods:

```
from __future__ import division
from sympy import, sin, pi

x=sin(50)

pi.evalf()
pi.evalf(50)

x.n()
x.n(20)
```

The next program demonstrates the concept of expressions and various operations that can be performed on expressions using the `collect`, `expand`, `factor`, `simplify`, and `subs` methods:

```
from sympy import collect, expand, factor, simplify
from sympy import Symbol, symbols
from sympy import sin, cos

x, y, a, b, c, d = symbols('x y a b c d')

expr = 5*x**2+2*b*x**2+cos(x)+51*x**2
simplify(expr)

factor(x**2+x-30)
expand ( (x-5) * (x+6) )

collect(x**3 + a*x**2 + b*x**2 + c*x + d, x)

expr = sin(x)*sin(x) + cos(x)*cos(x)
expr
expr.subs({x:5, y:25})
expr.subs({x:5, y:25}).n()
```

Equation solving

There is a magic function called `solve`. It can solve all types of equations. This function returns the solutions of an equation. It takes two arguments: the expression to be solved and the variable. The following program uses this function to solve various types of equations. In the following equations, it is assumed that the right-hand side of the equation is zero:

```
from sympy import solve

solve (6*x**2 - 3*x - 30,x)

a, b, c = symbols('a b c')
solve( a*x**2 + b*x + c, x)
substitute_solution = solve( a*x**2 + b*x + c, x)
[ substitute_solution[0].subs({'a':6,'b':-3,'c':-30}), substitute_
solution[1].subs({'a':6,'b':-3,'c':-30}) ]

solve([2*x + 3*y - 3, x -2* y + 1], [x, y])
)
```

To solve the system of equations, we have another form of the `solve` method that takes the list of equations as the first argument and the list of unknowns as the second argument. This is demonstrated here:

```
from sympy import solve

solve ([2*x + y - 4, 5*x - 3*y],[x, y])
solve ([2*x + 2*y - 1, 2*x - 4*y],[x, y])
```

Functions for rational numbers, exponentials, and logarithms

SymPy has a number of functions for working on rational numbers. These functions perform various operations on rational numbers, including simplify, expansion, combine, split, and many more. SymPy also supports several functions for exponential and logarithmic operations. There are three logarithm functions: `log` (used to compute base-b logarithms), `ln` (used to compute natural logarithms), and `log10` (used to compute base-10 logarithms). The `log` function expects two arguments: the variable and the base. If the base is not passed, then by default, this function will compute the natural logarithm of the variable, which is equivalent to `ln`. To calculate the addition of two rational numbers, we use the `together` function. Similarly, to divide a rational expression's numerator by a denominator, we use the `apart` function which is used in the following program:

```
from sympy import together, apart, symbols
x1, x2, x3, x4 = symbols('x1 x2 x3 x4')
x1/x2 + x3/x4
together(x1/x2 + x3/x4)

apart ((2*x**2+3*x+4)/(x+1))
together(apart ((2*x**2+3*x+4)/(x+1)))

exp(1)
log(4).n()
log(4,4).n()
ln(4).n()
mpmath.log10(4)
```

Polynomials

SymPy allows us to define and perform various operations on polynomials. We can also find the roots of polynomials. We have already covered the `simplify`, `expand`, `factor`, and `solve` methods. These methods also perform the functionalities for polynomials. To check for the equality of two polynomials, we should use the `simplify` function:

```
from sympy import *

p, q = symbols ('p q')
p = (x+4)*(x+2)
q = x**2 + 6*x + 8
p == q # Unsuccessful
p - q == 0 # Unsuccessful
simplify(p - q) == 0
```

Trigonometry and complex numbers

Mostly, the input for a trigonometric function is a radian angle, whereas an inverse trigonometric function returns the radian angle. This module also provides functions for conversion from degree to radian and radian to degree. Besides basic trigonometric functions, such as `sin`, `cos`, and `tan`, SymPy has trigonometry simplification and expansion functions.

SymPy also supports complex numbers to cope up with situations where no real number solution exists. For example, consider this equation: $x**2+4=0$. For this equation there is no real number solution; its solution will be -2*I or +2*I. This *I* denotes the square root of -1. The following program demonstrates trigonometric functions and gives a solution of this equation in the form of complex numbers:

```
from sympy import *
x, y = symbols('x y')
expr = sin(x)*cos(y)+cos(x)*sin(y)
expr_exp= exp(5*sin(x)**2+4*cos(x)**2)

trigsimp(expr)
trigsimp(expr_exp)
expand_trig(sin(x+y))
solve(x**2+4,x) #complex number as solution
```

Linear algebra

The SymPy linear algebra module is another very simple module that provides easy-to-learn functions for matrix manipulation. It has the functionality of performing various matrix operations, including quick special matrix creation, eigenvalues, eigenvectors, transpose, determinant, and inverse. There are three methods for quick special matrix creation, namely eye, zeros, and ones. The eye method creates an identity matrix, whereas zeros and ones create matrices with all elements equal to 0 or 1, respectively. If required, we can delete selected rows and columns from a matrix. Basic arithmetic operators, such as +, -, *, and **, also work on matrices:

```python
from sympy import *
A = Matrix( [[1, 2, 3, 4],
        [5, 6, 7, 8],
        [ 9, 10, 11, 12],
        [ 13, 14, 15, 16]] )
A.row_del(3)
A.col_del(3)

A[0,1] # display row 0, col 1 of A
A[0:2,0:3] # top-left submatrix(2x3)

B = Matrix ([[1, 2, 3],
        [5, 6, 7],
        [ 9, 10, 11]] )
A.row_join(B)
B.col_join(B)
A + B
A - B
A * B
A **2
eye(3) # 3x3 identity matrix
zeros(3, 3) # 3x3 matrix with all elements Zeros
ones(3, 3) # 3x3 matrix with all elements Ones

A.transpose() # It is same as A.T
M = Matrix( [[1, 2, 3],
    [4, 5, 6],
    [7, 8, 10]] )
M.det()
```

By default, the inverse of a matrix is computed by Gaussian elimination, and we can specify to have it computed by LU decomposition. Matrices in SymPy have methods for calculating the reduced row echelon form (the `rref` method) and null space (the `nullspace` method). If *A* is a matrix, then `nullspace` is the set of all vectors *v* such that *A v=0*. It is also possible to perform substitution operations on matrix elements; we can create a matrix with symbolic entries and substitute them with actual values and other symbols. We can also perform special operations, such as QR factorization, the Gram-Schmidt orthogonalizer, and LU decomposition:

```
from sympy import *
A = Matrix( [[1,2],
  [3,4]] )
A.inv()
A.inv()*A
A*A.inv()
A = Matrix( [[ 1, -2],
  [-2, 3]] )
A.eigenvals() # same as solve( det(A-eye(2)*x), x)
A.eigenvects()
A.rref()

A.nullspace()

x = Symbol('x')
M = eye(3) * x
M.subs(x, 4)
y = Symbol('y')
M.subs(x, y)

M.inv()
M.inv("LU")

A = Matrix([[1,2,1],[2,3,3],[1,3,2]])
Q, R = A.QRdecomposition()
Q

M = [Matrix([1,2,3]), Matrix([3,4,5]), Matrix([5,7,8])]
result1 = GramSchmidt(M)
result2 = GramSchmidt(M, True)
```

Calculus

Calculus involves operations that are performed to study the various properties of any function, including rates of change, the limit behavior of a function, and calculation of the area under a function graph. In this section, you will learn the concepts of limits, derivatives, summation of series, and integrals. The following program uses limit functions to solve simple limit problems:

```python
from sympy import limit, oo, symbols,exp, cos

oo+5
67000 < oo
10/oo

x , n = symbols ('x n')
limit( ((x**n - 1)/ (x - 1) ), x, 1)

limit( 1/x**2, x, 0)
limit( 1/x, x, 0, dir="-")

limit(cos(x)/x, x, 0)
limit(sin(x)**2/x, x, 0)
limit(exp(x)/x,x,oo)
```

Any SymPy expression can be differentiated using the `diff` function with the `diff(func_to_be_differentiated, variable)` prototype. The following program uses the `diff` function to compute the differentiation of various SymPy expressions:

```python
from sympy import diff, symbols, Symbol, exp, dsolve, subs, Function

diff(x**4, x)
diff( x**3*cos(x), x )
diff( cos(x**2), x )
diff( x/sin(x), x )
diff(x**3, x, 2)
diff( exp(x), x)
```

The `dsolve` method helps us solve any kind of ordinary differential equation and system of ordinary differential equations. This program demonstrates the use of `dsolve` for ordinary differential equations and boundary value problems:

```python
x = symbols('x')
f = symbols('f', cls=Function)
dsolve( f(x) - diff(f(x),x), f(x) )
```

```
#ics argument can be used to pass the boundary condition as a
dictionary to dsolve method
from sympy import *
x=symbols('x')
f=symbols('f', cls=Function)
dsolve(Eq(f(x).diff(x,x), 400*(f(x)-1+2*x)), f(x), ics={f(0):5,
f(2):10})
# the above line will results in f(x) = C1*e^-30x + C2*e^30x - 2x + 1
```

The following program finds critical points of the function *f(x)=4x3-3x2+2x* and uses the second derivative to find the maxima of the function in the interval *[0,1]*:

```
x = Symbol('x')
f = 4*x**3-3*x**2+2*x
diff(f, x)
sols = solve( diff(f,x), x)
sols
diff(diff(f,x), x).subs( {x:sols[0]} )
diff(diff(f,x), x).subs( {x:sols[1]} )
```

In SymPy, we can compute definite and indefinite integrals using the `integrate` function. Here is a program that computes definite and indefinite integrals. It will define these integrals symbolically. To compute the actual value, we call the `n()` method on the integral, as done in the last line of this program:

```
from sympy import *
integrate(x**3+1, x)
integrate(x*sin(x), x)
integrate(x+ln(x), x)

F = integrate(x**3+1, x)
F.subs({x:1}) - F.subs({x:0})

integrate(x**3-x**2+x, (x,0,1))     # definite Integrals
integrate(sin(x)/x, (x,0,pi))    # definite Integrals
integrate(sin(x)/x, (x,pi,2*pi))  # definite Integrals
integrate(x*sin(x)/(x+1), (x,0,2*pi)) # definite Integrals
integrate(x*sin(x)/(x+1), (x,0,2*pi)).n()
```

Sequences are functions that take integer numbers, and we can define a sequence by specifying an expression for its nth term. We can also substitute the desired value. The following program demonstrates the concept of sequences using some simple sequences in SymPy:

```
from sympy import *
s1_n = 1/n
s2_n = 1/factorial(n)
s1_n.subs({n:5})
[ s1_n.subs({n:index1}) for index1 in range(0,8) ]
[ s2_n.subs({n:index1}) for index1 in range(0,8) ]
limit(s1_n, n, oo)
limit(s2_n, n, oo)
```

A series whose terms contain different-ordered powers of a variable is called a power series, such as the Taylor series, exponential series, or sin/cos series. Here is a program that computes some sequences involving special functions. It also uses the concept of power series:

```
from sympy import *
s1_n = 1/n
s2_n = 1/factorial(n)
summation(s1_n, [n, 1, oo])
summation(s2_n, [n, 0, oo])
import math
def s2_nf(n):
  return 1.0/math.factorial(n)

sum( [s2_nf(n) for n in range(0,10)] )
E.evalf()

exponential_xn = x**n/factorial(n)
summation( exponential_xn.subs({x:5}), [x, 0, oo] ).evalf()
exp(5).evalf()
summation( exponential_xn.subs({x:5}), [x, 0, oo])

import math # redo using only python
def exponential_xnf(x,n):
  return x**n/math.factorial(n)
sum( [exponential_xnf(5.0,i) for i in range(0,35)] )

series( sin(x), x, 0, 8)
series( cos(x), x, 0, 8)
series( sinh(x), x, 0, 8)
series( cosh(x), x, 0, 8)
series(ln(x), x, 1, 6) # Taylor series of ln(x) at x=1
series(ln(x+1), x, 0, 6) # Maclaurin series of ln(x+1)
```

Vectors

An *n*-tuple defined on real numbers can also be called a vector. In physics and mathematics, a vector is a mathematical object that has either size, magnitude or length, and a direction. In SymPy, a vector is represented as a *1 x n* row matrix or an *n x 1* column matrix. The following program demonstrates the concept of vector computations in SymPy. It computes the transpose and norm of a vector:

```
from sympy import *
u = Matrix([[1,2,3]]) # a row vector = 1x3 matrix
v = Matrix([[4],
[5],    # a col vector = 3x1 matrix
[6]])
v.T # use the transpose operation to
u[1] # 0-based indexing for entries
u.norm() # length of u
uhat = u/u.norm() # unit-length vec in same dir as u
uhat
uhat.norm()
```

The next program demonstrates the concepts of dot product, cross product, and projection operations on vectors:

```
from sympy import *
u = Matrix([ 1,2,3])
v = Matrix([-2,3,3])
u.dot(v)

acos(u.dot(v)/(u.norm()*v.norm())).evalf()
u.dot(v) == v.dot(u)
u = Matrix([2,3,4])
n = Matrix([2,2,3])
(u.dot(n) / n.norm()**2)*n  # projection of v in the n dir

w = (u.dot(n) / n.norm()**2)*n
v = u - (u.dot(n)/n.norm()**2)*n # same as u - w
u = Matrix([ 1,2,3])
v = Matrix([-2,3,3])
u.cross(v)
(u.cross(v).norm()/(u.norm()*v.norm())).n()

u1,u2,u3 = symbols('u1:4')
v1,v2,v3 = symbols('v1:4')
Matrix([u1,u2,u3]).cross(Matrix([v1,v2,v3]))
u.cross(v)
v.cross(u)
```

The physics module

The physics module contains functionality required to solve the problem from physics. There are several submodules of physics for performing activities related to vector physics, classic mechanics, quantum mechanics, optics, and much more.

Hydrogen wave functions

There are two functions under this category. The first one computes the energy of state *(n, l)* in Hartree atomic units. The other computes the relativistic energy of state *(n, l, spin)* in Hartree atomic units. The following program demonstrates the use of these functions:

```
from sympy.physics.hydrogen import E_nl, E_nl_dirac, R_nl
from sympy import var

var("n Z")
var("r Z")
var("n l")
E_nl(n, Z)
E_nl(1)
E_nl(2, 4)

E_nl(n, l)
E_nl_dirac(5, 2) # l should be less than n
E_nl_dirac(2, 1)
E_nl_dirac(3, 2, False)
R_nl(5, 0, r) # z = 1 by default
R_nl(5, 0, r, 1)
```

Matrices and Pauli algebra

There are several matrices related to physics that are available in `physics.matrices` module. The following program demonstrates how to obtain these matrices and Pauli algebra:

```
from sympy.physics.paulialgebra import Pauli, evaluate_pauli_product
from sympy.physics.matrices import mdft, mgamma, msigma, pat_matrix

mdft(4) # expression of discrete Fourier transform as a matrix
multiplication
mgamma(2) # Dirac gamma matrix in the Dirac representation
msigma(2) #  Pauli matrix with (1,2,3)
```

```
# Following line computer Parallel Axis Theorem matrix to translate
the inertia matrix a distance of dx, dy, dz for a body of mass m.
pat_matrix(3, 1, 0, 0)

evaluate_pauli_product(4*x*Pauli(3)*Pauli(2))
```

The quantum harmonic oscillator in 1-D and 3-D

This module has functions for computation of energy of a one-dimensional harmonic oscillator, a three-dimensional isotropic harmonic oscillator, a wave function for a one-dimensional harmonic oscillator, and a radial wave function for a three-dimensional isotropic harmonic oscillator. Here is a program that uses the functions available in this module:

```
from sympy.physics.qho_1d import E_n, psi_n
from sympy.physics.sho import E_nl, R_nl
from sympy import var

var("x m omega")
var("r nu l")
x, y, z = symbols('x, y, z')

E_n(x, omega)
psi_n(2, x, m, omega)
E_nl(x, y, z)

R_nl(1, 0, 1, r)
R_nl(2, 1, 1, r)
```

Second quantization

The concept used to analyze and describe a quantum many-body system is called second quantization. This module contains classes for second quantization operators and states for bosons. Predefined symbols are available for import from `sympy.abc`:

```
from sympy.physics.secondquant import Dagger, B, Bd
from sympy.functions.special.tensor_functions import KroneckerDelta
from sympy.physics.secondquant import B, BKet, FockStateBosonKet
from sympy.abc import x, y, n
from sympy.abc import i, j, k
from sympy import Symbol
from sympy import I
```

```
Dagger(2*I)
Dagger(B(0))
Dagger(Bd(0))

KroneckerDelta(1, 2)
KroneckerDelta(3, 3)

#predefined symbols are available in abc including greek symbols like
theta
KroneckerDelta(i, j)
KroneckerDelta(i, i)
KroneckerDelta(i, i + 1)
KroneckerDelta(i, i + 1 + k)


a = Symbol('a', above_fermi=True)
i = Symbol('i', below_fermi=True)
p = Symbol('p')
q = Symbol('q')
KroneckerDelta(p, q).indices_contain_equal_information
KroneckerDelta(p, q+1).indices_contain_equal_information
KroneckerDelta(i, p).indices_contain_equal_information

KroneckerDelta(p, a).is_above_fermi
KroneckerDelta(p, i).is_above_fermi
KroneckerDelta(p, q).is_above_fermi


KroneckerDelta(p, a).is_only_above_fermi
KroneckerDelta(p, q).is_only_above_fermi
KroneckerDelta(p, i).is_only_above_fermi


B(x).apply_operator(y)

B(0).apply_operator(BKet((n,)))
sqrt(n)*FockStateBosonKet((n - 1,))
```

High-energy Physics

High-energy Physics is the study of the basic constituents of any matter and the associated forces. The following program demonstrates the use of the classes and functions defined in this module:

```
from sympy.physics.hep.gamma_matrices import GammaMatrixHead
from sympy.physics.hep.gamma_matrices import GammaMatrix,
DiracSpinorIndex
from sympy.physics.hep.gamma_matrices import GammaMatrix as GM
from sympy.tensor.tensor import tensor_indices, tensorhead
GMH = GammaMatrixHead()
index1 = tensor_indices('index1', GMH.LorentzIndex)
GMH(index1)


index1 = tensor_indices('index1', GM.LorentzIndex)
GM(index1)


GM.LorentzIndex.metric



p, q = tensorhead('p, q', [GMH.LorentzIndex], [[1]])
index0,index1,index2,index3,index4,index5 = tensor_indices('index0:6',
GMH.LorentzIndex)
ps = p(index0)*GMH(-index0)
qs = q(index0)*GMH(-index0)
GMH.gamma_trace(GM(index0)*GM(index1))
GMH.gamma_trace(ps*ps) - 4*p(index0)*p(-index0)
GMH.gamma_trace(ps*qs + ps*ps) - 4*p(index0)*p(-index0) -
4*p(index0)*q(-index0)


p, q = tensorhead('p, q', [GMH.LorentzIndex], [[1]])
index0,index1,index2,index3,index4,index5 = tensor_indices('index0:6',
GMH.LorentzIndex)
ps = p(index0)*GMH(-index0)
qs = q(index0)*GMH(-index0)
GMH.simplify_gpgp(ps*qs*qs)


index0,index1,index2,index3,index4,index5 = tensor_indices('index0:6',
GM.LorentzIndex)
spinorindex0,spinorindex1,spinorindex2,spinorindex3,spinorindex4,spin
orindex5,spinorindex6,spinorindex7 = tensor_indices('spinorindex0:8',
DiracSpinorIndex)
GM1 = GammaMatrix
```

```
t = GM1(index1,spinorindex1,-spinorindex2)*GM1(index4,spinorindex7,-sp
inorindex6)*GM1(index2,spinorindex2,-spinorindex3)*GM1(index3,spinorin
dex4,-spinorindex5)*GM1(index5,spinorindex6,-spinorindex7)
GM1.simplify_lines(t)
```

Mechanics

SymPy has a module that has the facilities and tools required for mechanical systems that are capable of manipulating reference frames, forces, and torques. The following program computes the net force acting on any object. The net force on an object is the sum of all the forces acting on that object. This is performed using vector addition, as the forces are vectors:

```
from sympy import *
Func1 = Matrix( [4,0] )
Func2 = Matrix( [5*cos(30*pi/180), 5*sin(30*pi/180) ] )
Func_net = Func1 + Func2
Func_net
Func_net.evalf()

Func_net.norm().evalf()
(atan2( Func_net[1],Func_net[0] )*180/pi).n()

t, a, vi, xi = symbols('t vi xi a')
v = vi + integrate(a, (t, 0,t) )
v
x = xi + integrate(v, (t, 0,t) )
x

(v*v).expand()
((v*v).expand() - 2*a*x).simplify()
```

If the net force on an object is constant, then the motion reflected by this constant force involves constant acceleration. The following program demonstrates this concept. It also uses the concept of **uniform-acceleration motion (UAM)**. In the previous program, the concept of potential energy is demonstrated:

```
From the sympy import *
xi = 20 # initial position
vi = 10 # initial velocity
a = 5 # acceleration (constant during motion)
x = xi + integrate( vi+integrate(a,(t,0,t)), (t,0,t) )
x
x.subs({t:3}).n() # x(3) in [m]
diff(x,t).subs({t:3}).n() # v(3) in [m/s]
```

```
t, vi, xi, k = symbols('t vi xi k')
a = sqrt(k*t)
x = xi + integrate( vi+integrate(a,(t,0,t)), (t, 0,t) )
x


x, y = symbols('x y')
m, g, k, h = symbols('m g k h')
F_g = -m*g # Force of gravity on mass m
U_g = - integrate( F_g, (y,0,h) )
U_g
F_s = -k*x # Spring force for displacement x
U_s = - integrate( F_s, (x,0,x) )
U_s
```

The next program uses the `dsolve` method to find the position function of the differential equation representation of the motion of a mass-spring system:

```
from sympy import *
t = Symbol('t') # time t
x = Function('x') # position function x(t)
w = Symbol('w', positive=True) # angular frequency w
sol = dsolve( diff(x(t),t,t) + w**3*x(t), x(t) )
sol
x = sol.rhs
x


A, phi = symbols("A phi")
(A*cos(w*t - phi)).expand(trig=True)


x = sol.rhs.subs({"C1":0,"C2":A})
x
v = diff(x, t)
E_T = (0.3*k*x**3 + 0.3*m*v**3).simplify()
E_T
E_T.subs({k:m*w**4}).simplify()
E_T.subs({w:sqrt(k/m)}).simplify()
```

Pretty printing

SymPy can pretty print the output using ASCII and Unicode characters. There are a number of printers available in SymPy. The following are the most common printers of SymPy:

- LaTeX
- MathML
- Unicode pretty printer
- ASCII pretty printer
- Str
- dot
- repr

This program demonstrates the pretty print function to print various expressions using the ASCII and Unicode printers:

```
from sympy.interactive import init_printing
from sympy import Symbol, sqrt
from sympy.abc import x, y
sqrt(21)
init_printing(pretty_print=True)
sqrt(21)
theta = Symbol('theta')
init_printing(use_unicode=True)
theta
init_printing(use_unicode=False)
theta
init_printing(order='lex')
str(2*y + 3*x + 2*y**2 + x**2+1)
init_printing(order='grlex')
str(2*y + 3*x + 2*y**2 + x**2+1)
init_printing(order='grevlex')
str(2*y * x**2 + 3*x * y**2)
init_printing(order='old')
str(2*y + 3*x + 2*y**2 + x**2+1)
init_printing(num_columns=10)
str(2*y + 3*x + 2*y**2 + x**2+1)
```

The following program uses the LaTeX printer for pretty printing. This is very useful when publishing the results of computation in a documentation or publication, which is a scientist's most general requirement:

```
from sympy.physics.vector import vprint, vlatex, ReferenceFrame,
dynamicsymbols

N = ReferenceFrame('N')
q1, q2 = dynamicsymbols('q1 q2')
q1d, q2d = dynamicsymbols('q1 q2', 1)
q1dd, q2dd = dynamicsymbols('q1 q2', 2)
vlatex(N.x + N.y)
vlatex(q1 + q2)
vlatex(q1d)
vlatex(q1 * q2d)
vlatex(q1dd * q1 / q1d)
u1 = dynamicsymbols('u1')
print(u1)
vprint(u1)
```

LaTeX Printing

LaTeX printing is implemented in the `LatexPrinter` class. It has a function for converting a given expression into a LaTeX representation. This program demonstrates the conversion of some mathematical expressions into LaTeX representations:

```
from sympy import latex, pi, sin, asin, Integral, Matrix, Rational
from sympy.abc import x, y, mu, r, tau

print(latex((2*tau)**Rational(15,4)))
print(latex((2*mu)**Rational(15,4), mode='plain'))
print(latex((2*tau)**Rational(15,4), mode='inline'))
print(latex((2*mu)**Rational(15,4), mode='equation*'))
print(latex((2*mu)**Rational(15,4), mode='equation'))
print(latex((2*mu)**Rational(15,4), mode='equation', itex=True))
print(latex((2*tau)**Rational(15,4), fold_frac_powers=True))
print(latex((2*tau)**sin(Rational(15,4))))
print(latex((2*tau)**sin(Rational(15,4)), fold_func_brackets = True))
print(latex(4*x**2/y))
print(latex(5*x**3/y, fold_short_frac=True))
print(latex(Integral(r, r)/3/pi, long_frac_ratio=2))
print(latex(Integral(r, r)/3/pi, long_frac_ratio=0))
print(latex((4*tau)**sin(Rational(15,4)), mul_symbol="times"))
```

```
print(latex(asin(Rational(15,4))))
print(latex(asin(Rational(15,4)), inv_trig_style="full"))
print(latex(asin(Rational(15,4)), inv_trig_style="power"))
print(latex(Matrix(2, 1, [x, y])))
print(latex(Matrix(2, 1, [x, y]), mat_str = "array"))
print(latex(Matrix(2, 1, [x, y]), mat_delim="("))
print(latex(x**2, symbol_names={x:'x_i'}))
print(latex([2/x, y], mode='inline'))
```

The cryptography module

This SymPy module includes methods for both block ciphers and stream ciphers. Specifically, it includes the following ciphers:

- Affine cipher
- Bifid cipher
- ElGamal encryption
- Hill's cipher
- Kid RSA
- Linear feedback shift registers
- RSA
- Shift cipher
- Substitution ciphers
- Vigenere's cipher

This program demonstrates the RSA deciphering and enciphering on plain text:

```
from sympy.crypto.crypto import rsa_private_key, rsa_public_key,
encipher_rsa, decipher_rsa
a, b, c = 11, 13, 17
rsa_private_key(a, b, c)
publickey = rsa_public_key(a, b, c)
pt = 8
encipher_rsa(pt, publickey)

privatekey = rsa_private_key(a, b, c)
ct = 112
decipher_rsa(ct, privatekey)
```

The following program performs Bifid cipher encryption and decryption on plain text and returns the cipher text:

```
from sympy.crypto.crypto import encipher_bifid6, decipher_bifid6
key = "encryptingit"
pt = "A very good book will be released in 2015"
encipher_bifid6(pt, key)
ct = "AENUIUKGHECNOIY27XVFPXR52XOXSPI0Q"
decipher_bifid6(ct, key)
```

Parsing input

The last module that we will be discussing is a small but useful module that parses input strings into SymPy expressions. Here is a program that demonstrates the use of this module. There are methods available for making parentheses optional, making multiplication implicit, and allowing functions to be instantiated:

```
from sympy.parsing.sympy_parser import parse_expr
from sympy.parsing.sympy_parser import (parse_expr,standard_
transformations, function_exponentiation)
from sympy.parsing.sympy_parser import (parse_expr,standard_
transformations, implicit_multiplication_application)

x = Symbol('x')
parse_expr("2*x**2+3*x+4"))

parse_expr("10*sin(x)**2 + 3xyz")

transformations = standard_transformations + (function_
exponentiation,)
parse_expr('10sin**2 x**2 + 3xyz + tan theta', transformations=transf
ormations)

parse_expr("5sin**2 x**2 + 6abc + sec theta",transformations=(standa
rd_transformations +(implicit_multiplication_application,)))
```

The logic module

The logic module allows users to create and manipulate logic expressions using symbolic and Boolean values. The user can build a Boolean expression using Python operators such as & (logical AND), | (logical OR), and ~ (logical NOT). The user can also create implications using >> and << . The following program demonstrates the use of these operators:

```
from sympy.logic import *
a, b = symbols('a b')
a | (a & b)
a | b
~a

a >> b
a << b
```

This module also has a function for logical Xor, Nand, Nor, logical implication, and the equivalence relation. These functions are used in the following program to demonstrate their capability. All of these functions support their symbolic forms and computations on these operators. In symbolic form, the expression represented in the symbol form, they are not evaluated. This is demonstrated using the a and b symbols:

```
from sympy.logic.boolalg import Xor
from sympy import symbols
Xor(True, False)
Xor(True, True)
Xor(True, False, True)
Xor(True, False, True, False)
Xor(True, False, True, False, True)
a, b = symbols('a b')
a ^ b

from sympy.logic.boolalg import Nand
Nand(True, False)
Nand(True, True)
Nand(a, b)

from sympy.logic.boolalg import Nor
Nor(True, False)
Nor(True, True)
Nor(False, True)
Nor(False, False)
Nor(a, b)
```

```
from sympy.logic.boolalg import Equivalent, And
Equivalent(False, False, False)
Equivalent(True, False, False)
Equivalent(a, And(a, True))

from sympy.logic.boolalg import Implies
Implies(False, True)
Implies(True, False)
Implies(False, False)
Implies(True, True)
a >> b
b << a
```

The logic module also allows users to use the if-then-else clause, convert a preposition logic sentence into a conjunctive or disjunctive normal form, and check whether or not an expression is in a conjunctive or disjunctive normal form. The following program demonstrates these functions. ITE returns the second argument if the first is true. Otherwise, it returns the third argument. The to_cnf and to_dnf functions perform a conversion of the expression or preposition statement into CNF and DNF, respectively; is_cnf and is_dnf confirm that the given expression is cnf and dnf, respectively:

```
from sympy.logic.boolalg import ITE, And, Xor, Or
from sympy.logic.boolalg import to_cnf, to_dnf
from sympy.logic.boolalg import is_cnf, is_dnf
from sympy.abc import A, B, C
from sympy.abc import X, Y, Z
from sympy.abc import a, b, c

ITE(True, False, True)
ITE(Or(True, False), And(True, True), Xor(True, True))
ITE(a, b, c)
ITE(True, a, b)
ITE(False, a, b)
ITE(a, b, c)

to_cnf(~(A | B) | C)
to_cnf((A | B) & (A | ~A), True)

to_dnf(Y & (X | Z))
to_dnf((X & Y) | (X & ~Y) | (Y & Z) | (~Y & Z), True)

is_cnf(X | Y | Z)
is_cnf(X & Y & Z)
is_cnf((X & Y) | Z)
```

```
is_cnf(X & (Y | Z))

is_dnf(X | Y | Z)
is_dnf(X & Y & Z)
is_dnf((X & Y) | Z)
is_dnf(X & (Y | Z))
```

The logic module has a `simplify` method that converts a Boolean expression into its simplified **sum of product (SOP)** or **product of sum (POS)** form. There are functions that use the simplified pair and redundant group elimination algorithm, which converts all input combinations that generate 1 to the smallest SOP or POS form. The following program demonstrates the use of these functions:

```
from sympy.logic import simplify_logic
from sympy.logic import SOPform, POSform
from sympy.abc import x, y, z
from sympy import S

minterms = [[0, 0, 0, 1], [0, 0, 1, 1], [0, 1, 1, 1], [1, 0, 1, 1],
[1, 1, 1, 1]]
dontcares = [[1, 1, 0, 1], [0, 0, 0, 0], [0, 0, 1, 0]]
SOPform(['w','x','y','z'], minterms, dontcares)

minterms = [[0, 0, 0, 1], [0, 0, 1, 1], [0, 1, 1, 1], [1, 0, 1, 1],
[1, 1, 1, 1]]
dontcares = [[1, 1, 0, 1], [0, 0, 0, 0], [0, 0, 1, 0]]
POSform(['w','x','y','z'], minterms, dontcares)

expr = '(~x & y & ~z) | ( ~x & ~y & ~z)'
simplify_logic(expr)
S(expr)
simplify_logic(_)
```

The geometry module

The geometry module allows creation, manipulation, and computations on two-dimensional shapes, including points, lines, circles, ellipses, polygons, triangles, and others. The next program demonstrates the creation of these shapes and some operations on the `collinear` function. This function tests whether a given set of points is collinear, and it returns true if they are collinear. Points are called collinear if they lie on a single straight line. The `medians` function returns a dictionary with a vertex as the key, and the value is the median at that vertex. The `intersection` function finds the intersection points of two or more geometrical entities. Whether a given line is a tangent to a circle or not is determined by the `is_tangent` method.

The `circumference` function returns the circumference of a circle, and the `equation` function returns the circle in its equation form:

```
from sympy import *
from sympy.geometry import *

x = Point(0, 0)
y = Point(1, 1)
z = Point(2, 2)
zp = Point(1, 0)

Point.is_collinear(x, y, z)
Point.is_collinear(x, y, zp)

t = Triangle(zp, y, x)
t.area
t.medians[x]

Segment(Point(1, S(1)/2), Point(0, 0))
m = t.medians
intersection(m[x], m[y], m[zp])

c = Circle(x, 5)
l = Line(Point(5, -5), Point(5, 5))
c.is_tangent(l)
l = Line(x, y)
c.is_tangent(l)
intersection(c, l)

c1 = Circle( Point(2,2), 7)
c1.circumference()
c1.equation()
l1 = Line (Point (0,0), Point(10,10))
intersection (c1,l1)
```

The geometry module has several specialized submodules for performing operations on various two-dimensional and some three-dimensional shapes. The following are the submodules of this module:

- **Points**: This represents a point in two-dimensional Euclidean space.

- **3D Point**: This class represents a point in three-dimensional Euclidean space.

- **Lines**: This represents an infinite 2D line in space.

- **3D Line**: This represents an infinite 3D line in space.

- **Curves**: This represents a curve in space. A curve is an object similar to a line, but it is not required to be straight.

- **Ellipses**: This class represents an elliptical geometry entity.

- **Polygon**: This represents a two-dimensional polygon. A polygon is a figure that is a closed circuit or chain bounded by a finite number of line segments. These line segments are called edges or sides of the polygon, and the connecting points of two edges are called vertices of the polygon.

- **Plane**: This represents a geometric plane, which is a flat two-dimensional surface. A plane can be considered a 2D analogue of a point in zero dimensions, a line in one dimension, and a solid in three-dimensional space.

Symbolic integrals

Methods meant for calculating definite and indefinite integrals of a given expression are implemented in the integrals module. There are two main methods in this module—one for definite integrals and other for indefinite integrals—as follows:

- **Integrate(f , x)**: This computes the indefinite integral of function f with respect to x ($\int f dx$)

- **Integrate(f, (x, m, n))**: This computes the definite integral of f with respect to x in the limit m to n ($\int mn f dx$)

This module allows users to compute integrals on various types of functions, ranging from simple polynomials to complex exponential polynomials. The following program performs integration on a number of functions to demonstrate its capability:

```
from sympy import integrate, log, exp, oo
from sympy.abc import n, x, y
from sympy import sqrt
from sympy import *
integrate(x*y, x)
integrate(log(x), x)
integrate(log(x), (x, 1, n))
integrate(x)
integrate(sqrt(1 + x), (x, 0, x))
integrate(sqrt(1 + x), x)
integrate(x*y)
integrate(x**n*exp(-x), (x, 0, oo)) # same as conds='piecewise'
integrate(x**n*exp(-x), (x, 0, oo), conds='none')
integrate(x**n*exp(-x), (x, 0, oo), conds='separate')
```

```
init_printing(use_unicode=False, wrap_line=False, no_global=True)
x = Symbol('x')
integrate(x**3 + x**2 + 1, x)
integrate(x/(x**3+3*x+1), x)
integrate(x**3 * exp(x) * cos(x), x)
integrate(exp(-x**3)*erf(x), x)
```

This module also has the following advance functions for the computation of points as weights for various quadratures of different orders and precision. Furthermore, it has several special functions for definite integrals and various integral transforms.

The numerical integral in the quadrature submodule (`sympy.integrals.quadrature`) contains the functions used to perform computations on the following quadratures:

- The Gauss-Legendre quadrature
- The Gauss-Laguerre quadrature
- The Gauss-Hermite quadrature
- Gauss-Chebyshev quadrature
- Gauss-Jacobi quadrature

Integral transforms contains methods for the following transforms submodules in the transform module (`sympy.integrals.transforms`):

- Mellin Transform
- Inverse Mellin Transform
- Laplace Transform
- Inverse Laplace Transform
- Unitary ordinary-frequency Fourier Transform
- Unitary ordinary-frequency inverse Fourier Transform
- Unitary ordinary-frequency sine Transform
- Unitary ordinary-frequency inverse sine Transform
- Unitary ordinary-frequency cosine Transform
- Unitary ordinary-frequency inverse cosine Transform
- Hankel Transform

Polynomial manipulation

The Polys module in SymPy allows users to perform polynomial manipulations. It has methods ranging from simple operations on polynomials, such as division, GCD, and LCM, to advanced concepts, such as Gröbner bases and multivariate factorization.

The following program shows polynomial division using the div method. This method performs polynomial division with the remainder. An argument domain may be used to specify the types of values of the argument. If the operation is to be performed only on integers, then pass domain='ZZ', domain='QQ' for rational and domain='RR' for real numbers. The expand method expands the expression into its normal representation:

```
from sympy import *
x, y, z = symbols('x,y,z')
init_printing(use_unicode=False, wrap_line=False, no_global=True)

f = 4*x**2 + 8*x + 5
g = 3*x + 1
q, r = div(f, g, domain='QQ')   ## QQ for rationals
q
r
(q*g + r).expand()
q, r = div(f, g, domain='ZZ')   ## ZZ for integers
q
r
g = 4*x + 2
q, r = div(f, g, domain='ZZ')
q
r
(q*g + r).expand()
g = 5*x + 1
q, r = div(f, g, domain='ZZ')
q
r
(q*g + r).expand()
a, b, c = symbols('a,b,c')
f = a*x**2 + b*x + c
g = 3*x + 2
q, r = div(f, g, domain='QQ')
q
r
```

The following program demonstrates LCM, GCD, square-free factorization, and simple factorization. Square-free factorization is performed using the `sqf` method. The SQF of a univariate polynomial is the product of all factors of degree 1 and 2. On the other hand, the `factor` method performs factorization of univariate and multivariate polynomials of rational coefficients:

```
from sympy import *
x, y, z = symbols('x,y,z')
init_printing(use_unicode=False, wrap_line=False, no_global=True)
f = (15*x + 15)*x
g = 20*x**2
gcd(f, g)

f = 4*x**2/2
g = 16*x/4
gcd(f, g)

f = x*y/3 + y**2
g = 4*x + 9*y
gcd(f, g)

f = x*y**2 + x**2*y
g = x**2*y**2
gcd(f, g)

lcm(f, g)
(f*g).expand()
(gcd(f, g, x, y)*lcm(f, g, x, y)).expand()

f = 4*x**2 + 6*x**3 + 3*x**4 + 2*x**5
sqf_list(f)
sqf(f)

factor(x**4/3 + 6*x**3/16 - 2*x**2/4)
factor(x**2 + 3*x*y + 4*y**2)
```

Sets

The Sets SymPy module enables users to perform set theory computations. It has classes, or submodules, for representing various types of sets, such as a finite set (a finite set of discrete numbers) and an interval (represents a real interval as a set), a singleton set, a universal set, naturals (sets of natural numbers), and others. It also has submodules for performing various operations on compound sets, such as union, intersection, product set, complement, and others.

The following program demonstrates the creation of an interval set and a finite set. It also demonstrates the start and end attributes of the interval set and left open and right open interval sets. At the end, the program also uses the option of checking the existence of a specific element in a finite set:

```
from sympy import Symbol, Interval
from sympy import FiniteSet

Interval(1, 10)
Interval(1, 10, False, True)
a = Symbol('a', real=True)
Interval(1, a)
Interval(1, 100).end
from sympy import Interval
Interval(0, 1).start

Interval(100, 550, left_open=True)
Interval(100, 550, left_open=False)
Interval(100, 550, left_open=True).left_open
Interval(100, 550, left_open=False).left_open

Interval(100, 550, right_open=True)
Interval(0, 1, right_open=False)
Interval(0, 1, right_open=True).right_open
Interval(0, 1, right_open=False).right_open

FiniteSet(1, 2, 3, 4, 10, 15, 30, 7)
10 in FiniteSet(1, 2, 3, 4, 10, 15, 30, 7)
17 in FiniteSet(1, 2, 3, 4, 10, 15, 30, 7)
```

The next program demonstrates operations on compound sets, such as union, intersection, product of sets, and complement. A union of two sets will be a set that has all the elements from both the sets. On the other hand, an intersection of sets results in a new set that has only those elements that are common in the given sets. A product set represents the Cartesian product of the given sets. A complement of sets represents the set difference or relative complement of one set with respect to another:

```
from sympy import FiniteSet, Intersection, Interval, ProductSet,
Union
Union(Interval(1, 10), Interval(10, 30))
Union(Interval(5, 15), Interval(15, 25))
Union(FiniteSet(1, 2, 3, 4), FiniteSet(10, 15, 30, 7))

Intersection(Interval(1, 3), Interval(2, 4))
Interval(1,3).intersect(Interval(2,4))
Intersection(FiniteSet(1, 2, 3, 4), FiniteSet(1, 3, 4, 7))
FiniteSet(1, 2, 3, 4).intersect(FiniteSet(1, 3, 4, 7))

I = Interval(0, 5)
S = FiniteSet(1, 2, 3)
ProductSet(I, S)
(2, 2) in ProductSet(I, S)

Interval(0, 1) * Interval(0, 1)
coin = FiniteSet('H', 'T')
set(coin**2)

Complement(FiniteSet(0, 1, 2, 3, 4, 5), FiniteSet(1, 2))
```

The simplify and collect operations

The SymPy module also supports the simplify and collect operations on the given expression. There are options for simplifying various types of functions, including trigonometric functions, Bessel-type functions, combinatorial expressions, and others.

The following program demonstrates the simplification of expressions involving polynomial and trigonometric functions:

```
from sympy import simplify, cos, sin, trigsimp, cancel
from sympy import sqrt, count_ops, oo, symbols, log
from sympy.abc import x, y
```

```
expr = (2*x + 3*x**2)/(4*x*sin(y)**2 + 2*x*cos(y)**2)
expr
simplify(expr)

trigsimp(expr)
cancel(_)

root = 4/(sqrt(2)+3)
simplify(root, ratio=1) == root
count_ops(simplify(root, ratio=oo)) > count_ops(root)
x, y = symbols('x y', positive=True)
expr2 = log(x) + log(y) + log(x)*log(1/y)

expr3 = simplify(expr2)
expr3
count_ops(expr2)
count_ops(expr3)
print(count_ops(expr2, visual=True))
print(count_ops(expr3, visual=True))
```

Summary

In this chapter, we extensively discussed computing on a computerized algebra system. We also saw symbol creation, the use of expressions, and basic arithmetic. Then we discussed equation solvers and covered functions for rational numbers, exponentials, and logarithms. We also discussed the functionality for polynomials, trigonometry, and complex numbers.

Topics such as linear algebra, calculus, vectors, and concepts related to physics were covered in the later part of the chapter. Finally, we discussed pretty printing, cryptography, and string parsing into an expression.

In the next chapter, we will have an exhaustive discussion on Python visual computing using matplotlib and pandas. We will cover how to visualize data and the results of computations. Using pandas, we will also cover data analysis for scientific computing.

7
Data Analysis and Visualization

In this chapter, we will be discussing the concepts of data visualization, plotting, and interactive computing using matplotlib, pandas, and IPython. Data visualization is the process of presenting data in graphic or pictorial form. This will help understand information from data easily and quickly. By "plotting," we mean representing the dataset in the form of a graph to show the relationship between two or more variables. By "interactive computing," we mean the software that accepts input from the user. Generally, these are commands to be processed by the software. After accepting the input, the software performs processing as per the command entered by the user. These concepts will be accompanied by example programs.

In this chapter, we will be covering the following topics:

- Concepts associated with plotting, using matplotlib
- Types of the plots, using sample programs
- The fundamental concepts of pandas
- The pandas structures, using sample programs
- Performing data analysis activities using pandas
- The components of interactive computing, using IPython
- Using the various components of IPython

pandas is a library that has high-performance and easy-to-use data structures and data analysis tools. It allows users to draw plots of various types with standard and customized styles.

IPython is a command shell for interactive computing in multiple programming languages. It was specially designed for Python.

Matplotlib

The most popular Python package for working on two-dimensional graphics and chart plotting is matplotlib. It provides a very quick way of data visualization in the form of different types of plots/charts. It also supports exporting of these plots into various formats. We will be starting the discussion of matplotlib with the basics and architecture, and then we will discuss the plotting of various types of charts using sample programs.

The architecture of matplotlib

The most important matplotlib object is `Figure`. It contains and manages all the elements of the given charts/graphics. matplotlib has separated the figure representation and manipulation activity from the rendering of `Figure` to the user interface screen or the devices. This enables users to design and develop interesting features and logic, while the backend and device manipulation remains very simple. It supports graphics rendering for multiple devices and also supports event handling of popular user interface designing toolkits.

The matplotlib architecture is separated into three layers, namely backend, artist, and scripting. These three layers form a stack, wherein each upper layer knows the way of communication with the lower layer but the lower layer is not aware of the upper layer. The backend layer is the bottommost layer, the scripting layer is the topmost layer, and the artist layer is the middle layer. Now let's discuss the details of these layers from top to bottom.

The scripting layer (pyplot)

The `pyplot` interface of matplotlib is intuitive and very simple to use by scientists and analysts. It simplifies the common tasks to be performed for analysis and visualization. The `pyplot` interface manages the activities of creating figures, axes, and their connection with the backend. It hides the internal details of the maintenance of data structures to represent the figures and axis.

Let's discuss a sample program that demonstrates the simplicity of this layer:

```
import matplotlib.pyplot as plt
import numpy as np
var = np.random.randn(5300)
plt.hist(var, 530)
plt.title(r'Normal distribution ($\mu=0, \sigma=1$)')
plt.show()
```

To save the histogram in an image file, we can add the `plt.savefig('sample_histogram.png')` quotes text as the last but one line to the preceding code just before showing it.

The artist layer

This middle layer of the matplotlib stack handles most of the internal activities behind the great plots. The base class of this layer is `matplotlib.artist.Artist`. This object knows how to use the renderer to draw on the canvas. Each of the displayed objects on the matplotlib `Figure` is an instance of `Artist`, including the titles, axis and data labels, image, lines, bars, and points. An individual `Artist` instance is created for each of these components.

There are a number of attributes associated with the artist shared by each instance. The first attribute is transformation, which performs the translation of artist coordinates into the canvas coordinate system. The next attribute is visibility. It is the region where the artist can do the drawing. The labels in the drawing are also an attribute, and the final attribute is an interface that handles user activities performed by a mouse-like click.

The backend layer

The bottommost layer is the backend layer, which has an actual implementation of a number of abstract interface classes, namely `FigureCanvas`, `Renderer`, and `Event`. `FigureCanvas` is the class that plays the concept of the surface used to draw. In an analogy with real painting, `FigureCanvas` is equivalent to the paper used in painting. `Renderer` plays the role of the painting component, which is performed by the paintbrush in real-life painting. The `Event` class handles the keyboard and mouse events.

This layer also supports integration with user interface toolkits, such as Qt. The abstract base classes for integration with these user interface toolkits reside in `matplotlib.backend_bases`. The classes derived for specific user interface toolkits are kept in dedicated modules, such as `matplotlib.backends.backend_qt4agg`.

For creating an image, the backend layer has headers, fonts, and functions meant to store the output in files of different formats, including PDF, PNG, PS, SVG, and so on.

The `Renderer` class provides the drawing interface that actually performs the drawing on the canvas.

Graphics with matplotlib

Using matplotlib, a user can draw a variety of two-dimensional plots. This section covers some simple plots and two special types of plots: contour and vector plots. The following program is for drawing a line plot on the radius and area of a circle:

```
import matplotlib.pyplot as plt
#radius
r = [1.5, 2.0, 3.5, 4.0, 5.5, 6.0]
#area of circle
a = [7.06858, 12.56637, 38.48447, 50.26544, 95.03309, 113.09724]
plt.plot(r, a)
plt.xlabel('Radius')
plt.ylabel('Area')
plt.title('Area of Circle')
plt.show()
```

The next program is for drawing a line plot that has two different lines so as to represent sine and cosine lines. Generally, these types of plots are used for comparison. There are various choices for colors, line styles, and markers. The third argument to the plot method represents the line color, line style, and marker. The first character represents the colors, which can have any value among b, g, r, c, m, y, k, and w. Here, others are obvious and k represents black. The second and the following characters indicate the line type, which can take any of these values: -, --, -.., and :. These symbols indicate solid, dashed, dash dotted, and dotted lines, respectively. The last character indicates data markers are ., x, +, o, and *:

```
import matplotlib.pyplot as plt
var = arange(0.,100,0.2)
cos_var = cos(var)
sin_var = sin(var)
plt.plot(var,cos_var,'b-*',label='cosine')
plt.plot(var,sin_var,'r-.',label='sine')
plt.legend(loc='upper left')
plt.xlabel('xaxis')
plt.ylabel('yaxis')
plt.show()
```

In the graph, we can set the limitation for the *x* and *y* axes using the xlim or ylim function. Try to add plot.ylim(-2,2) as the last but one line in the preceding program, and observe the impact.

The following program is for generating a histogram plot on Gaussian numbers. These numbers are generated using normal method:

```
import matplotlib.pyplot as plt
from numpy.random import normal
sample_gauss = normal(size=530)
plt.hist(sample_gauss, bins=15)
plt.title("Histogram Representing Gaussian Numbers")
plt.xlabel("Value")
plt.ylabel("Frequency")
plt.show()
```

The next program will generate a contour plot on the linearly spaced vector generated using the `linspace` method on a defined function:

```
import matplotlib.pyplot as plt
from numpy import *
x = linspace(0,10.5,40)
y = linspace(1,8,30)
(X,Y) = meshgrid(x,y)
func = exp(-((X-2.5)**2 + (Y-4)**2)/4) - exp(-((X-7.5)**2 + (Y-
4)**2)/4)
contr = plt.contour(x,y,func)
plt.clabel(contr)
plt.xlabel("x")
plt.ylabel("y")
plt.show()
```

The following program generates a vector plot, again on the linearly spaced vectors generated using the `linspace` method. We can store the graph elements in variables if we need to reuse them later in any form. This is shown in the second and third lines from the bottom in the following program, which store `xlabel` and `ylabel` in variables:

```
import matplotlib.pyplot as plt
from numpy import *
x = linspace(0,15,11)
y = linspace(0,10,13)
(X,Y) = meshgrid(x,y)
arr1 = 15*X
arr2 = 15*Y
main_plot = plt.quiver(X,Y,arr1,arr2,angles='xy',scale=1000,color='b')
main_plot_key = plt.quiverkey(main_plot,0,15,30,"30 m/s",coordinates='
data',color='b')
xl = plt.xlabel("x in (km)")
yl = plt.ylabel("y in (km)")
plt.show()
```

Output generation

The output of the plotting that is generated is a graph, which can be saved in different formats, including images, PDF, and PS. To store the output in a file, we have two options:

- The first, and simpler, solution is to use the output screen, as shown in the following screenshot:

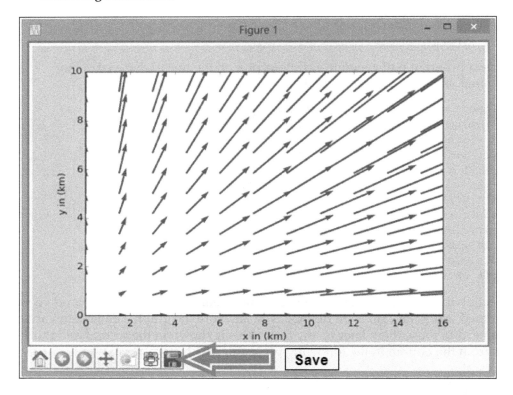

On the output screen, in the bottom-left corner, there are many buttons, out of which the rightmost button can be used to save the figure in a file. This will open a dialog box that tells you to save the file. Save that file in an appropriate folder with a desired type and a specified name.

- The second method is to save the figure in a file using the `plt.savefig` method just before the `plt.show()` method. We can also specify the filename and file format / type using this method.

The following program stores multiple figures in a single PDF file on different pages. It also demonstrates some techniques for saving the figure in a PNG image file:

```python
from matplotlib.backends.backend_pdf import PdfPages
import matplotlib.pyplot as plt
import matplotlib as mpl
from numpy.random import normal
from numpy import *

# PDF initialization
pdf = mpl.backends.backend_pdf.PdfPages("output.pdf")

# First Plot as first page of the PDF
sample_gauss = normal(size=530)
plt.hist(sample_gauss, bins=15)
plt.xlabel("Value")
plt.ylabel("Frequency")
plt.title("Histogram Representing Gaussian Numbers")
pdf.savefig()
plt.close()

# create second plot and saved on second page of PDF
var = arange(0.,100,0.2)
cos_var = cos(var)
sin_var = sin(var)
plt.legend(loc='upper left')
plt.xlabel('xaxis')
plt.ylabel('yaxis')
plt.plot(var,cos_var,'b-*',label='cosine')
plt.plot(var,sin_var,'r-.',label='sine')
pdf.savefig()
pdf.close()
plt.close()

# output to a PNG file
r = [1.5, 2.0, 3.5, 4.0, 5.5, 6.0]
a = [7.06858, 12.56637, 38.48447, 50.26544, 95.03309, 113.09724]
plt.plot(r, a)
plt.xlabel('Radius')
plt.ylabel('Area')
plt.title('Area of Circle')
plt.savefig("sample_output.png")
plt.show()
```

The pandas library

The pandas library has tools that support high-performance data analysis tasks. This library is useful for both commercial and scientific applications. The acronym "pandas" is partially derived from the econometric term "panel data" and Python data analysis. The five typical steps of data analysis and data processing are load, prepare, manipulate, model, and analyze.

pandas has added three new data structures to Python, namely Series, DataFrame, and Panel. These data structures are developed on top of NumPy. Let's discuss each of these data structures in detail.

Series

Series is a one-dimensional object similar to an array, a list, or a column in a table. It can hold any of Python's data types, including integers, floating-point numbers, strings, and any Python object. It also assigns a labeled index to each item in a series. By default, it will assign labels from *0* to *N* to a series that have *N-1* items. We can create a Series using the `Series` method, from an ndarrays, or from the dictionary (`dict`). Ideally, we should also pass the indices along with the data in the series.

Let's discuss the use of the Series data structure in a sample program:

```
import numpy as np
randn = np.random.randn
from pandas import *

s = Series(randn(10), index=['I', 'II', 'III', 'IV', 'V', 'VI', 'VII',
'VIII', 'IX', 'X' ])
s
s.index

Series(randn(10))

d = {'a' : 0., 'e' : 1., 'i' : 2.}
Series(d)
Series(d, index=['e', 'i', 'o', 'a'])

#Series creation using scalar value
Series(6., index=['a', 'e', 'i', 'o', 'u', 'y'])
Series([10, 20, 30, 40], index=['a', 'e', 'i', 'o'])
Series({'a': 10, 'e': 20, 'i': 30})
```

```
s.get('VI')

# name attribute can be specified
s = Series(np.random.randn(5), name='RandomSeries')
```

DataFrame

pandas's two-dimensional data structure is called **DataFrame**. A DataFrame is a data structure that is composed of rows and columns, similar to database tables or spreadsheets.

Similar to the series, a DataFrame also accepts a variety of input, such as these:

- Dictionary of one-dimensional ndarrays, list, series, and `dict`.
- Two-dimensional ndarrays
- The ndarrays of a structure/record
- A Series or a DataFrame

Although the index and column arguments are optional, it is better to pass them. The index can be referred to as row labels, and columns can be referred to as column labels. The following program first creates the DataFrame from `dict`. If the column names are not passed, then it means that the column names are the sorted key values.

After this, the program also creates a DataFrame from `dict` of ndarrays/list. Finally, it creates the DataFrame from the array of the structure or record:

```
import numpy as np
randn = np.random.randn
from pandas import *

#From Dict of Series/ dicts
d = {'first' : Series([10., 20., 30.], index=['I', 'II', 'III']),
     'second' : Series([10., 20., 30., 40.], index=['I', 'II', 'III',
'IV'])}
DataFrame(d, index=['IV', 'II', 'I'])

DataFrame(d, index=['IV', 'II', 'I'], columns=['second', 'third'])
df = DataFrame(d)
df
df.index
df.columns
```

```
#dict of ndarray/list
d = {'one' : [10., 20., 30., 40.],
     'two' : [40., 30., 20., 10.]}
DataFrame(d)
DataFrame(d, index=['I', 'II', 'III', 'IV'])

# Array of Structure/ record
data = np.zeros((2,),dtype=[('I', 'i4'),('II', 'f4'),('III', 'a10')])
data[:] = [(10,20.,'Very'),(20,30.,"Good")]

DataFrame(data)
DataFrame(data, index=['first', 'second'])
DataFrame(data, columns=['III', 'I', 'II'])
```

Panel

The Panel data structure is useful for storing three-dimensional data. The term is derived from statistics and econometrics, where multidimensional data contains measurements over a time period. Generally, the Panel data contains observations of multiple data items over different periods of time for the same organization or persons.

There are three main components of a panel — item, major axis, and minor axis — as explained here:

- `items`: Items represents data items of DataFrame inside the Panel
- `major_axis`: This represents the indexes (row labels) of DataFrames
- `minor_axis`: This represents the columns of the DataFrames

The following program demonstrates various methods of creating a Panel: item selection/indexing, squeezing, and conversion to a hierarchical indexed DataFrame. The last two lines of this program convert the Panel into the DataFrame using the `to_frame` method:

```
import numpy as np
randn = np.random.randn
from pandas import *

# Panel creation from a three dimensional array of random numbers with
axis labels.
workpanel = Panel(randn(2, 3, 5), items=['FirstItem', 'SecondItem'],
    major_axis=date_range('1/1/2010', periods=3),
    minor_axis=['A', 'B', 'C', 'D', 'E'])
workpanel
```

```
# Panel creation from Dict of DataFrame
data = {'FirstItem' : DataFrame(randn(4, 3)),
        'SecondItem' : DataFrame(randn(4, 2))}
Panel(data)

# orient=minor indicates to use the DataFrame's column as items
Panel.from_dict(data, orient='minor')

df = DataFrame({'x': ['one', 'two', 'three', 'four'],'y': np.random.
randn(4)})
df

data = {'firstitem': df, 'seconditem': df}
panel = Panel.from_dict(data, orient='minor')
panel['x']
panel['y']
panel['y'].dtypes

#Select a particular Item
workpanel['FirstItem']

# To rearrange the panel we can use transpose method.
workpanel.transpose(2, 0, 1)




# Fetch a slice at given major_axis label
workpanel.major_xs(workpanel.major_axis[1])

workpanel.minor_axis
# Fetch a slice at given minor_axis label
workpanel.minor_xs('D')

# The dimensionality can be changes using squeeze method.
workpanel.reindex(items=['FirstItem']).squeeze()
workpanel.reindex(items=['FirstItem'],minor=['B']).squeeze()



forconversionpanel = Panel(randn(2, 4, 5), items=['FirstItem',
'SecondItem'],
      major_axis=date_range('1/1/2010', periods=4),
      minor_axis=['A', 'B', 'C', 'D', 'E'])
forconversionpanel.to_frame()
```

The common functionality among the data structures

There are certain common functionalities among these data structures. These functions perform the same operations on these data structures. There are some common attributes among the data structures. The following program demonstrates the common functions/operations and attributes of pandas's data structures:

```
import numpy as np
randn = np.random.randn
from pandas import *

index = date_range('1/1/2000', periods=10)

s = Series(randn(10), index=['I', 'II', 'III', 'IV', 'V', 'VI', 'VII',
'VIII', 'IX', 'X' ])

df = DataFrame(randn(10, 4), index=['I', 'II', 'III', 'IV', 'V', 'VI',
'VII', 'VIII', 'IX', 'X' ], columns=['A', 'B', 'C', 'D'])

workpanel = Panel(randn(2, 3, 5), items=['FirstItem', 'SecondItem'],
    major_axis=date_range('1/1/2010', periods=3),
    minor_axis=['A', 'B', 'C', 'D', 'E'])

series_with100elements = Series(randn(100))

series_with100elements.head()
series_with100elements.tail(3)

series_with100elements[:3]
df[:2]
workpanel[:2]

df.columns = [x.lower() for x in df.columns]
df

# Values property can be used to access the actual value.
s.values
df.values
wp.values
```

There are some functions/attributes that can be performed/used on Series and DataFrame only. This program demonstrates the use of such functions and attributes, including describe, the min/max index, sorting by labels/actual values, conversion of object functions, and dtypes attributes:

```python
import numpy as np
randn = np.random.randn
from pandas import *

# Describe Function
series = Series(randn(440))
series[20:440] = np.nan
series[10:20]   = 5
series.nunique()
series = Series(randn(1700))
series[::3] = np.nan
series.describe()

frame = DataFrame(randn(1200, 5), columns=['a', 'e', 'i', 'o', 'u'])
frame.ix[::3] = np.nan
frame.describe()

series.describe(percentiles=[.05, .25, .75, .95])
s = Series(['x', 'x', 'y', 'y', 'x', 'x', np.nan, 'u', 'v', 'x'])
s.describe()

frame = DataFrame({'x': ['Y', 'Yes', 'Yes', 'N', 'No', 'No'], 'y':
range(6)})
frame.describe()
frame.describe(include=['object'])
frame.describe(include=['number'])
frame.describe(include='all')


# Index min and max value
s1 = Series(randn(10))
s1
s1.idxmin(), s1.idxmax()

df1 = DataFrame(randn(5,3), columns=['X','Y','Z'])
df1
df1.idxmin(axis=0)
df1.idxmax(axis=1)
```

```
df3 = DataFrame([1, 2, 2, 3, np.nan], columns=['X'],
index=list('aeiou'))
df3
df3['X'].idxmin()

# sorting by label and sorting by actual values
unsorted_df = df.reindex(index=['a', 'e', 'i', 'o'],
                columns=['X', 'Y', 'Z'])
unsorted_df.sort_index()
unsorted_df.sort_index(ascending=False)
unsorted_df.sort_index(axis=1)

df1 = DataFrame({'X':[5,3,4,4],'Y':[5,7,6,8],'Z':[9,8,7,6]})
df1.sort_index(by='Y')
df1[['X', 'Y', 'Z']].sort_index(by=['X','Y'])

s = Series(['X', 'Y', 'Z', 'XxYy', 'Yxzx', np.nan, 'ZXYX', 'Zoo',
'Yet'])
s[3] = np.nan
s.order()
s.order(na_position='first')

# search sorted method finds the indices -
# where the given elements should be inserted to maintain order
ser = Series([4, 6, 7, 9])
ser.searchsorted([0, 5])
ser.searchsorted([1, 8])
ser.searchsorted([5, 10], side='right')
ser.searchsorted([1, 8], side='left')

s = Series(np.random.permutation(17))
s
s.order()
s.nsmallest(5)
s.nlargest(5)

# we can sort on multiple index
df1.columns = MultiIndex.from_tuples([('x','X'),('y','Y'),('z','X')])
df1.sort_index(by=('x','X'))

# Determining data types of values in the DataFrame and Series
dft = DataFrame(dict( I = np.random.rand(5),
```

```
                        II = 8,
                        III = 'Dummy',
                        IV = Timestamp('19751008'),
                        V = Series([1.6]*5).astype('float32'),
                        VI = True,
                        VII = Series([2]*5,dtype='int8'),
                VIII = False))
dft
dft.dtypes
dft['III'].dtype
dft['II'].dtype

# counts the occurrence of each data type
dft.get_dtype_counts()

df1 = DataFrame(randn(10, 2), columns = ['X', 'Y'], dtype = 'float32')
df1
df1.dtypes

df2 = DataFrame(dict( X = Series(randn(10)),
                      Y = Series(randn(10),dtype='uint8'),
                      Z = Series(np.array(randn(10),dtype='float16'))
))
df2
df2.dtypes

#Object conversion on DataFrame and Series
df3['D'] = '1.'
df3['E'] = '1'
df3.convert_objects(convert_numeric=True).dtypes
# same, but specific dtype conversion
df3['D'] = df3['D'].astype('float16')
df3['E'] = df3['E'].astype('int32')
df3.dtypes

s = Series([datetime(2001,1,1,0,0),
            'foo', 1.0, 1, Timestamp('20010104'),
            '20010105'],dtype='O')
s
s.convert_objects(convert_dates='coerce')
```

Performing iterations is very simple, and it works in the same way on all the data structures. There is an accessor for performing date operations on the Series data structure. The following program demonstrates these concepts:

```python
import numpy as np
randn = np.random.randn
from pandas import *

workpanel = Panel(randn(2, 3, 5), items=['FirstItem', 'SecondItem'],
    major_axis=date_range('1/1/2010', periods=3),
    minor_axis=['A', 'B', 'C', 'D', 'E'])
df = DataFrame({'one-1' : Series(randn(3), index=['a', 'b', 'c']),
                'two-2' : Series(randn(4), index=['a', 'b', 'c',
'd']),
        'three-3' : Series(randn(3), index=['b', 'c', 'd'])})

for columns in df:
    print(columns)


for items, frames in workpanel.iteritems():
    print(items)
    print(frames)

for r_index, rows in df2.iterrows():
        print('%s\n%s' % (r_index, rows))

df2 = DataFrame({'x': [1, 2, 3, 4, 5], 'y': [6, 7, 8, 9, 10]})
print(df2)
print(df2.T)

df2_t = DataFrame(dict((index,vals) for index, vals in df2.
iterrows()))
print(df2_t)

df_iter = DataFrame([[1, 2.0, 3]], columns=['x', 'y', 'z'])
row = next(df_iter.iterrows())[1]

print(row['x'].dtype)
print(df_iter['x'].dtype)

for row in df2.itertuples():
    print(row)
```

```
# datetime handling using dt accessor
s = Series(date_range('20150509 01:02:03',periods=5))
s
s.dt.hour
s.dt.second
s.dt.day
s[s.dt.day==2]

# Timezone based translation can be performed very easily
stimezone = s.dt.tz_localize('US/Eastern')
stimezone
stimezone.dt.tz
s.dt.tz_localize('UTC').dt.tz_convert('US/Eastern')

# period
s = Series(period_range('20150509',periods=5,freq='D'))
s
s.dt.year
s.dt.day

# timedelta
s = Series(timedelta_range('1 day 00:00:05',periods=4,freq='s'))
s
s.dt.days
s.dt.seconds
s.dt.components
```

pandas provides a large number of methods to perform computations of descriptive statistics and aggregation functions, such as count, sum, minimum, maximum, mean, median, mode, standard deviation, variance, skewness, kurtosis, quantile, and cumulative functions.

The following program demonstrates the use of these functions on the Series, DataFrame, and Panel data structures. These methods have an optional attribute name called skipna that specifies whether to exclude the missing data (NaN). By default, this argument is True:

```
import numpy as np
randn = np.random.randn
from pandas import *

df = DataFrame({'one-1' : Series(randn(3), index=['a', 'b', 'c']),
                'two-2' : Series(randn(4), index=['a', 'b', 'c',
'd']),
```

```
          'three-3' : Series(randn(3), index=['b', 'c', 'd'])})
df
df.mean(0)
df.mean(1)
df.mean(0, skipna=False)
df.mean(axis=1, skipna=True)
df.sum(0)
df.sum(axis=1)
df.sum(0, skipna=False)
df.sum(axis=1, skipna=True)

# the NumPy methods excludes missing values
np.mean(df['one-1'])
np.mean(df['one-1'].values)

ser = Series(randn(10))
ser.pct_change(periods=3)

# Percentage change over a given period
df = DataFrame(randn(8, 4))
df.pct_change(periods=2)

ser1 = Series(randn(530))
ser2 = Series(randn(530))
ser1.cov(ser2)

frame = DataFrame(randn(530, 5), columns=['i', 'ii', 'iii', 'iv',
'v'])
frame.cov()
frame = DataFrame(randn(26, 3), columns=['x', 'y', 'z'])
frame.ix[:8, 'i'] = np.nan
frame.ix[8:12, 'ii'] = np.nan
frame.cov()
frame.cov(min_periods=10)
frame = DataFrame(randn(530, 5), columns=['i', 'ii', 'iii', 'iv',
'v'])
frame.ix[::4] = np.nan

# By pearson (Default) method Standard correlation coefficient
frame['i'].corr(frame['ii'])
# We can specify method Kendall/ spearman
frame['i'].corr(frame['ii'], method='kendall')
frame['i'].corr(frame['ii'], method='spearman')

index = ['i', 'ii', 'iii', 'iv']
columns = ['first', 'second', 'third']
```

```
df1 = DataFrame(randn(4, 3), index=index, columns=columns)
df2 = DataFrame(randn(3, 3), index=index[:3], columns=columns)
df1.corrwith(df2)
df2.corrwith(df1, 1)

s = Series(np.random.randn(10), index=list('abcdefghij'))
s['d'] = s['b'] # so there's a tie
s.rank()

df = DataFrame(np.random.randn(8, 5))
df[4] = df[2][:5] # some ties
df
df.rank(1)
```

Time series and date functions

pandas has a range of time series and date manipulation functions that can be used to perform computations that require calculations of time and date.

There are a number of components that can be accessed from TimeStamp data. The following is the list of selected components:

- **year**: The year of the datetime
- **month**: The month of the datetime
- **day**: The days of the datetime
- **hour**: The hour of the datetime
- **minute**: The minutes of the datetime
- **second**: The seconds of the datetime
- **microsecond**: The microseconds of the datetime
- **nanosecond**: The nanoseconds of the datetime
- **date**: Returns datetime date
- **time**: Returns datetime time
- **dayofyear**: The ordinal day of year
- **weekofyear**: The week ordinal of the year
- **dayofweek**: The day of the week with Monday=0 and Sunday=6
- **quarter**: Quarter of the date with Jan-Mar=1, Apr-Jun=2, and so on.

Here is a program that demonstrates these functions:

```
import numpy as np
randn = np.random.randn
from pandas import *
# Date Range creation, 152 hours from 06/03/2015
range_date = date_range('6/3/2015', periods=152, freq='H')
range_date[:5]

# Indexing on the basis of date
ts = Series(randn(len(range_date)), index= range_date)
ts.head()

#change the frequency to 40 Minutes
converted = ts.asfreq('40Min', method='pad')
converted.head()
ts.resample('D', how='mean')
dates = [datetime(2015, 6, 10), datetime(2015, 6, 11), datetime(2015,
6, 12)]
ts = Series(np.random.randn(3), dates)
type(ts.index)
ts

#creation of period index
periods = PeriodIndex([Period('2015-10'), Period('2015-11'),
                       Period('2015-12')])
ts = Series(np.random.randn(3), periods)
type(ts.index)
ts

# Conversion to Timestamp
to_datetime(Series(['Jul 31, 2014', '2015-01-08', None]))
to_datetime(['1995/10/31', '2005.11.30'])
# dayfirst to represent the data starts with day
to_datetime(['01-01-2015 11:30'], dayfirst=True)
to_datetime(['14-03-2007', '03-14-2007'], dayfirst=True)
# Invalid data can be converted to NaT using coerce=True
to_datetime(['2012-07-11', 'xyz'])
to_datetime(['2012-07-11', 'xyz'], coerce=True)

#doesn't works properly on mixed datatypes
to_datetime([1, '1'])
# Epoch timestamp : Integer and float epoch times can be converted to
timestamp
```

```
# the default using is Nanoseconds that can be changed to seconds/
microseconds
# The base time is 01/01/1970
to_datetime([1449720105, 1449806505, 1449892905,
             1449979305, 1450065705], unit='s')
to_datetime([1349720105100, 1349720105200, 1349720105300,
             1349720105400, 1349720105500 ], unit='ms')
to_datetime([8])
to_datetime([8, 4.41], unit='s')

#Datetime Range
dates = [datetime(2015, 4, 10), datetime(2015, 4, 11), datetime(2015,
4, 12)]
index = DatetimeIndex(dates)
index = Index(dates)
index = date_range('2010-1-1', periods=1700, freq='M')
index
index = bdate_range('2014-10-1', periods=250)
index

start = datetime(2005, 1, 1)
end = datetime(2015, 1, 1)
range1 = date_range(start, end)
range1
range2 = bdate_range(start, end)
range2
```

Datetime information can also be used for indexing in a data structure. The following program demonstrates the use of datetime as an index. It also demonstrates the use of the DateOffset object:

```
import numpy as np
randn = np.random.randn
from pandas import *
from pandas.tseries.offsets import *

start = datetime(2005, 1, 1)
end = datetime(2015, 1, 1)
rng = date_range(start, end, freq='BM')
ts = Series(randn(len(rng)), index=rng)
ts.index
ts[:8].index
ts[::1].index
```

```
# We can directly use the dates and Strings for index
ts['8/31/2012']
ts[datetime(2012, 07, 11):]
ts['10/08/2005':'12/31/2014']
ts['2012']
ts['2012-7']

dft = DataFrame(randn(50000,1),columns=['X'],index=date_range('2005010
1',periods=50000,freq='T'))
dft
dft['2005']
# first time of the first month and last time of month in parameter
after :
dft['2005-1':'2013-4']
dft['2005-1':'2005-3-31']
# We can specify stop time
dft['2005-1':'2005-3-31 00:00:00']
dft['2005-1-17':'2005-1-17 05:30:00']
#Datetime indexing
dft[datetime(2005, 1, 1):datetime(2005,3,31)]
dft[datetime(2005, 1, 1, 01, 02, 0):datetime(2005, 3, 31, 01, 02, 0)]

#selection of single row using loc
dft.loc['2005-1-17 05:30:00']
# time trucation
ts.truncate(before='1/1/2010', after='12/31/2012')
```

Handling missing data

By missing data, we mean data that is null or not present, for any reason. Generally, it is represented as Na*, where * represents a single character, such as N for number (NaN) and T for DateTimes (NaT). The next program demonstrates a pandas function meant for checking missing data such as isNull and notNull, and filling in missing data using fillna, dropna, loc, iloc, and interpolate. If we perform any operation on NaN, it will result in NaN:

```
import numpy as np
randn = np.random.randn
from pandas import *

df = DataFrame(randn(8, 4), index=['I', 'II', 'III', 'IV', 'VI',
'VII', 'VIII', 'X' ],
    columns=['A', 'B', 'C', 'D'])
```

```
df['E'] = 'Dummy'
df['F'] = df['A'] > 0.5
df

# Introducing some Missing data by adding new index
df2 = df.reindex(['I', 'II', 'III', 'IV', 'V', 'VI', 'VII', 'VIII',
'IX', 'X'])
df2
df2['A']
#Checking for missing values
isnull(df2['A'])
df2['D'].notnull()

df3 = df.copy()
df3['timestamp'] = Timestamp('20120711')
df3
# Observe the output of timestamp column for missing values as NaT
df3.ix[['I','III','VIII'],['A','timestamp']] = np.nan
df3

s = Series([5,6,7,8,9])
s.loc[0] = None
s

s = Series(["A", "B", "C", "D", "E"])
s.loc[0] = None
s.loc[1] = np.nan
s

# Fillna method to fill the missing value
df2
df2.fillna(0)  # fill all missing value with 0
df2['D'].fillna('missing') # fill particular column missing value with
missing

df2.fillna(method='pad')
df2
df2.fillna(method='pad', limit=1)

df2.dropna(axis=0)
df2.dropna(axis=1)
```

```
ts
ts.count()
ts[10:30]=None
ts.count()
# interpolate method perform interpolation to fill the missing values
# By default it performs linear interpolation
ts.interpolate()
ts.interpolate().count()
```

I/O operations

The pandas I/O API is a bundle of reader functions that returns a pandas object. It is very easy to load data using the tools bundled in pandas. Data is loaded into the pandas data structures from records in various types of files, such as **comma-separated values** (**CSV**), Excel, HDF, SQL, JSON, HTML, Google Big Query, pickle, stats format, and the clipboard. There are several reader functions—one function for each type of file—namely read_csv, read_excel, read_hdf, read_sql, read_json, read_html, read_stata, read_clipboard, and read_pickle. After loading, the data is prepared for analyzing. This involves deletion of erroneous entries, normalization, grouping, transformation, and sorting.

Working on CSV files

The next program demonstrates working on CSV files and performing various operations on it. This program uses Book-Crossing datasets in CSV format, downloaded from http://www2.informatik.uni-freiburg.de/~cziegler/BX/. It contains three CSV files (BX-Books.csv, BX-Users.csv, and BX-Book-Ratings.csv). These have the details of books, users, and ratings given to the books by users. There are two options for passing the filename of CSV; we can either put the file in any folder and use the full path, or keep the file in the current directory and pass only its name. The file path in the following program is the full path on the Windows operating system:

```
import numpy as np
randn = np.random.randn
from pandas import *

user_columns = ['User-ID', 'Location', 'Age']
users = read_csv('c:\BX-Users.csv', sep=';', names=user_columns)

rating_columns = ['User-ID', 'ISBN', 'Rating']
ratings = read_csv('c:\BX-Book-Ratings.csv', sep=';', names=rating_columns)
```

```
book_columns = ['ISBN', 'Book-Title', 'Book-Author', 'Year-Of-
Publication', 'Publisher', 'Image-URL-S']
books = read_csv('c:\BX-Books.csv', sep=';', names=book_columns,
usecols=range(6))

books
books.dtypes

users.describe()
print books.head(10)
print books.tail(8)
print books[5:10]

users['Location'].head()
print users[['Age', 'Location']].head()

desired_columns = ['User-ID', 'Age']
print users[desired_columns].head()

print users[users.Age > 25].head(4)
print users[(users.Age < 50) & (users.Location == 'chicago, illinois,
usa')].head(4)

print users.set_index('User-ID').head()
print users.head()

with_new_index = users.set_index('User-ID')
print with_new_index.head()
users.set_index('User_ID', inplace=True)
print users.head()

print users.ix[62]
print users.ix[[1, 100, 200]]
users.reset_index(inplace=True)
print users.head()
```

Here is a program that demonstrates merge, groupby, and related operations, such as sorting, ordering, finding the top *n* values, and aggregation on the Book-Crossing datasets:

```
import numpy as np
randn = np.random.randn
from pandas import *
```

```
user_columns = ['User-ID', 'Location', 'Age']
users = read_csv('c:\BX-Users.csv', sep=';', names=user_columns)
rating_columns = ['User-ID', 'ISBN', 'Rating']
ratings = read_csv('c:\BX-Book-Ratings.csv', sep=';', names=rating_
columns)

book_columns = ['ISBN', 'Title', 'Book-Author', 'Year-Of-Publication',
'Publisher', 'Image-URL-S']
books = read_csv('c:\BX-Books.csv', sep=';', names=book_columns,
usecols=range(6))

# create one merged DataFrame
book_ratings = merge(books, ratings)
users_ratings = merge(book_ratings, users)

most_rated = users_ratings.groupby('Title').size().
order(ascending=False)[:25]
print most_rated

users_ratings.Title.value_counts()[:17]

book_stats = users_ratings.groupby('Title').agg({'Rating': [np.size,
np.mean]})
print book_stats.head()

# sort by rating average
print book_stats.sort([('Rating', 'mean')], ascending=False).head()

greater_than_100 = book_stats['Rating'].size >= 100
print book_stats[greater_than_100].sort([('Rating', 'mean')],
ascending=False)[:15]

top_fifty = users_ratings.groupby('ISBN').size().
order(ascending=False)[:50]
```

 The following program works on the CSV file available at https://
github.com/gjreda/gregreda.com/blob/master/content/
notebooks/data/city-of-chicago-salaries.csv?raw=true.

This program demonstrates the merging and concatenation operations on DataFrame:

```
import numpy as np
randn = np.random.randn
from pandas import *

first_frame = DataFrame({'key': range(10),
                         'left_value': ['A', 'B', 'C', 'D', 'E',
'F', 'G', 'H', 'I', 'J']})
second_frame = DataFrame({'key': range(2, 12),
                          'right_value': ['L', 'M', 'N', 'O', 'P',
'Q', 'R', 'S', 'T', 'U']})
print first_frame
print second_frame

#Natural Join Operation
print merge(left_frame, right_frame, on='key', how='inner')
# Left, Right and Full Outer Join Operation
print merge(left_frame, right_frame, on='key', how='left')
print merge(left_frame, right_frame, on='key', how='right')
print merge(left_frame, right_frame, on='key', how='outer')

concat([left_frame, right_frame])
concat([left_frame, right_frame], axis=1)

headers = ['name', 'title', 'department', 'salary']
chicago_details = read_csv('c:\city-of-chicago-salaries.csv',
                    header=False,
                    names=headers,
                    converters={'salary': lambda x: float(x.
replace('$', ''))})
print chicago_detail.head()

dept_group = chicago_details.groupby('department')

print dept_group
print dept_group.count().head(10)
print dept_group.size().tail(10)
print dept_group.sum()[10:17]
print dept_group.mean()[10:17]
print dept_group.median()[10:17]

chicago_details.sort('salary', ascending=False, inplace=True)
```

Ready-to-eat datasets

There are various sources of data on economics and specific modules for using this data in pandas programs. We can use the `pandas.io.data` and `pandas.io.ga` (Google Analytics) modules to extract data from various Internet sources and add it to DataFrame. At present, it supports the following sources:

- Yahoo! Finance
- Google Finance
- St. Louis Fed: **Federal reserve economic data** (FRED) is a database of over 267,000 economic time series from 80 sources
- Kenneth French's data library
- World Bank
- Google Analytics

Here is a small program that demonstrates the reading of data from some of these data sources:

```
import pandas.io.data as web
import datetime
f1=web.DataReader("F", 'yahoo', datetime.datetime(2010, 1, 1),
datetime.datetime(2011, 12, 31))
f2=web.DataReader("F", 'google', datetime.datetime(2010, 1, 1),
datetime.datetime(2011, 12, 31))
f3=web.DataReader("GDP", "fred", datetime.datetime(2010, 1, 1),
datetime.datetime(2011, 12, 31))
f1.ix['2010-05-12']
```

The pandas plotting

The pandas data structures support plot methods wrapper around the `plt.plot()` method for plotting of data in data structures. By default, it will display the line plot, which can be changed by passing an optional attribute named as kind to the plot method. The following list contains the changes in `df.plot()` for producing different plots:

- **Bar plot**: `df.plot(kind='bar')`
- **Histogram**: `df.plot(kind='hist')`
- **Box plot**: `df.plot(kind='box')`
- **Area plot**: `df.plot(kind='area')` -
- **Scatterplot**: `df.plot(kind='scatter')`
- **Pie plot**: `df.plot(kind='pie')`

This program demonstrates a simple plotting example from the pandas wrapper method. The output of the program is displayed in the screenshot shown after it:

```
from pandas import *
randn = np.random.randn
import matplotlib.pyplot as plt
x1 = np.array( ((1,2,3), (1,4,6), (2,4,8)) )
df = DataFrame(x1, index=['I', 'II', 'III'], columns=['A', 'B', 'C'])
df = df.cumsum()
df.plot(kind='pie', subplots=True)
plt.figure()
plt.show()
```

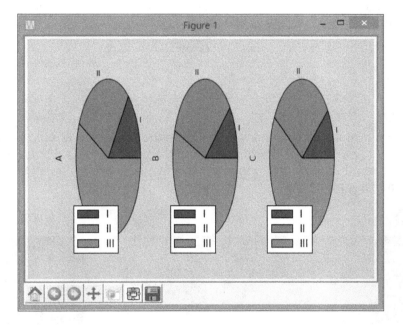

IPython

IPython is designed and developed with the aim of providing an enhanced Python shell that makes it possible to perform interactive distributed and parallel computing. IPython also has a set of tools for building special-purpose interactive environments for scientific computing. It has two components that help fulfill the aim of IPython:

- An enhanced interactive IPython shell
- Architecture for interactive parallel computing

In this section, we will first discuss the components that enhance the interactive IPython shell. We will cover the other component for interactive parallel computing in *Chapter 8, Parallel and Large-scale Scientific Computing.*

The IPython console and system shell

The interface provided by IPython is shown in the next screenshot. We can apply a different coloring scheme to this console; the default coloring scheme is NoColor. We have other options such as Linux and LightBG. An important feature of IPython is that it is stateful, as it maintains the state of the computations performed on the console. The output of any step in IPython is stored in _N, where N is the number of outputs/results. When we enter the IPython interactive shell, it displays the facility offered by this enhanced interactive IPython, as follows:

```
IPython 3.0.0 -- An enhanced Interactive Python.
?          -> Introduction and overview of IPython's features.
%quickref -> Quick reference.
help       -> Python's own help system.
object?    -> Details about 'object', use 'object??' for extra details.
```

If we enter a question mark (?) as a command in the shell, then it will display a detailed list of the features of IPython. Similarly, %quickref will display a short reference of a number of IPython commands, and %magic will display the details of IPython magic commands.

If we type any `objectname?`, then the console will display all the details about that object, such as the docstring, functions, and constructors, as depicted in the following screenshot. We have created a DataFrame object named `df` and displayed its details using `df?`.

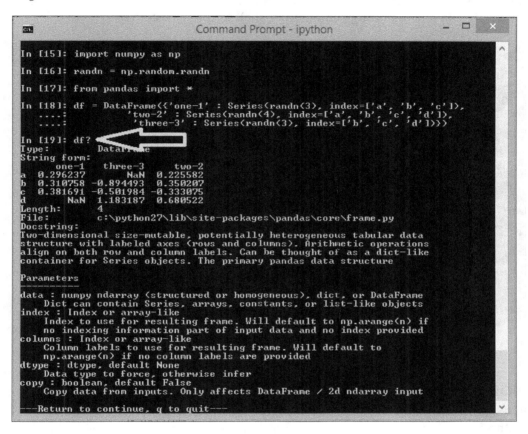

The operating system interface

There are a number of situations where it is desired to perform a computation with support from the operating system. The user can create their new aliases for frequently used commands. It also supports Unix commands such as `ls`. The user can prefix `!` to any operating command or shell script to execute it.

An operating system command executed inside Python shel

Nonblocking plotting

In the normal Python shell, if we create any plot and display it using the `show()` method, then the plot will be displayed on a new screen, and this keeps the shell blocked until the user closes the screen showing plot. However, IPython has a flag called `-pylab`. If we execute the IPython shell using the `IPython -pylab` command, then the plot windows that open from the IPython shell will not block the shell. This is presented in the following screenshot—a plot window opened without blocking the shell, as IPython is executed with the `-pylab` flag:

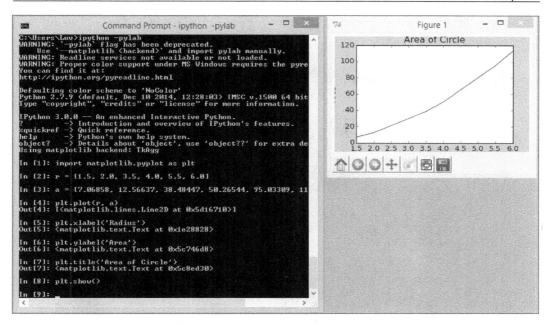

Debugging

IPython has excellent support for the debugging of programs and tracing of errors and exceptions. After the execution of the script, we can call %debug to start the Python debugger (pdb) to examine the problem. We can perform debugging activities here, as we can print the values of variables, execute statements, track the source of a specific problem. Generally, this avoids the use of external debugger applications.

This screenshot depicts the `%debug` option:

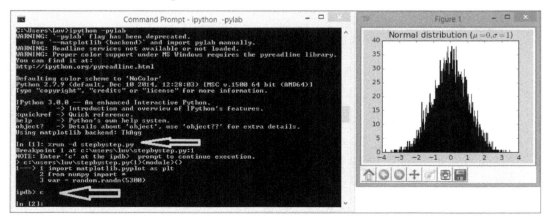

The user can execute any program step by step by calling `%run -d programname.py`. This is presented in the following screenshot. We have a step in the program named `stepbystep.py`. At each breakpoint, the debugger interface asks the user to press C to continue to the next step:

IPython Notebook

IPython has a web-based application called Notebook. It is designed and developed for interactive development and authoring of literate computations, where the text for explaining the concept, mathematical aspects, actual computations, and graphical output can be combined. The input to a program and the output of a program are stored in cells that may be edited in-place, if required.

The following screenshot, which is taken from `http://ipython.org/notebook.html`, presents an interface of IPython:

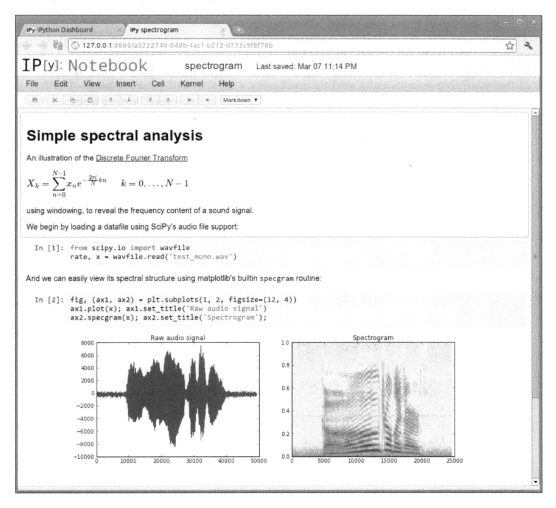

Summary

In this chapter, we started off by discussing the basic concepts and architecture of matplotlib. After that, we discussed some sample programs used to generate different types of plots. We also covered the methods of saving these plots in files of different formats. Then we discussed the use of pandas in data analysis.

Furthermore, we discussed the data structures of pandas. After covering the uses of data structures in depth, you learned how to perform various other related activities for data analysis. In the last part, we discussed the concepts, uses, and applications of interactive computing using IPython.

In the next chapter, we will have a comprehensive discussion on using Python for scientific computing that involves parallel and high-performance computing. The chapter will cover the basic concepts of parallel and high-performance computing, and the available frameworks and technologies. Later, it will provide an in-depth coverage of the use of Python for parallel and high-performance computing.

8

Parallel and Large-scale Scientific Computing

This chapter discusses the important concept of using parallel and large-scale computing in Python, or using IPython to solve scientific computing problems. It covers recent trends in large-scale scientific computing and Big Data processing. We will use example programs to understand these concepts.

In this chapter, we will cover the following topics:

- The fundamentals of parallel computing in IPython
- The components of IPython parallel computing
- IPython's task interface and database
- IPython's direct interface
- Details of IPython parallel computing
- The MPI program in IPython
- Big data processing using Hadoop and Spark in Python

IPython runs a number of different processes to enable users to perform parallel computing. The first of these processes is the IPython engine, which is a Python interpreter that executes the tasks submitted by users. A user can run multiple engines to perform parallel computation. The second process is the IPython hub, which monitors the engines and schedulers to keep track of the status of user tasks. The hub process listens for registration requests from engines and clients; it continuously monitors connections from schedulers. The third process is the IPython scheduler. This is a group of processes used to communicate the commands and results between clients and engines. Generally, the scheduler process runs on the machine that runs the controller process and is connected to the hub process. The last process is the IPython client, which is the IPython session that is used to coordinate the engines to perform parallel computations.

All of these processes are collectively called as IPython cluster. These processes use ZeroMQ for communication with each other. ZeroMQ supports various transport protocols including Infiband, IPC, PGM, TCP, and others. The IPython controller, which is composed of a hub and schedulers, listens to the clients' requests on sockets. When the user starts an engine, it connects to a hub and performs the registration. Now the hub exchanges the connection information of the schedulers with the engine. Later, the engine connects to the schedulers. These connections persist throughout the life of each engine. Each IPython client uses many socket connections to connect to the controller. Generally, it uses one connection per scheduler and three connections for a hub. These connections are maintained throughout the life of the client.

Parallel computing using IPython

IPython allows users to perform parallel and high-performance computing in an interactive manner. We can use IPython's built-in support for parallel computing, which consists of four components that make IPython suitable for most types of parallelism. Specifically, IPython supports the following types of parallelism:

- **Single program, multiple data parallelism (SPMD)**: This is the most common style of parallel programming, and it is a subtype of **Multiple Instruction and Multiple Data (MIMD)**. In this model, each task executes its own copy of the same program. Each task processes different datasets to achieve better performance.

- **Multiple program, multiple data parallelism**: In the **multiple program, multiple data (MPMD)** style, each task executes different programs that process different datasets on each participant computing node.

- **Message passing using MPI**: A **Message Passing Interface (MPI)** is a specification for developers of message passing libraries. It is a language-independent specification that enables its users to write message-passing-based parallel programs. In its present form, it supports distributed shared memory models and their hybrid models.

- **Task parallelism**: Task parallelism distributes execution processes among the different nodes involved in computing. The task can be threads, a message-passing component, or a component of some other programming model, such as MapReduce.

- **Data parallelism**: Data parallelism distributes data across the different nodes involved in parallel computing. The main emphasis of data parallelism is distribution/parallelization of data across different nodes, in contrast to task parallelization.

- **Hybrid parallelism of the aforementioned types**: IPython also supports any parallel computing style that is a hybrid of any of the aforementioned styles.

- **User-defined approaches to parallelism**: IPython is designed to be very simple and highly flexible, and this designing focus enables users to use it for any new or user-defined parallelism style.

IPython supports the various phases of the program development life cycle for all types of parallel applications in an interactive style, such as development, execution, debugging, monitoring, and so on.

Using matplotlib with IPython enables users to analyze and visualize remote or distributed large datasets. It also enables them to start job processing on a cluster and pull back the data on the local system for analysis and visualization. Users can push MPI applications onto a high-performance computer from an IPython session on a desktop/laptop. It also supports dynamic load balancing for different tasks running on a set of CPUs. Furthermore, it supports a number of simple approaches that allow users to interactively parallelize many simple applications in two or three lines of code. Users can interactively develop, execute, test, and debug custom parallel algorithms. IPython enables users to bundle different MPI jobs in execution on different computing nodes into single, huge distributed and/or parallel system.

The architecture of IPython parallel computing

The architecture of parallel computing in IPython has three main components. These components are part of the parallel package of IPython. The architecture of IPython parallel computing is depicted in the following figure:

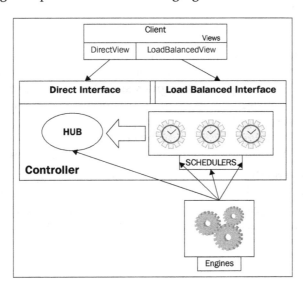

The three main components of IPython parallel computing are **client**, **controller**, and **engines**. The **controller** component is composed of two subcomponents: **HUB** and **SCHEDULERS**. It allows client interaction with engines through two main interfaces: direct interface and load-balanced interface.

The components of parallel computing

Various components and concepts related to the IPython parallel computing architecture will be discussed in this subsection. The components are the IPython engine, the IPython controller (the hub and schedulers), and the IPython clients and views.

The IPython engine

The core component performs the actual execution of the Python command received as a network request. The engine is an instance of a regular Python interpreter, and ultimately it will become an entire IPython interpreter. A user can perform distributed computing and parallel computing by starting multiple engines. The user code is executed in the IPython engine in blocking mode.

The IPython controller

The controller is composed of a hub and a group of schedulers. An IPython controller is a bundle of processes used for communication by clients and engines. It is the point of contact for users who have Python processes to be executed by engines. Generally, the schedulers are separate processes that run on the same machine on which the hub runs. In some special cases, schedulers run on a remote machine:

- **Hub**: The hub is the most important component, and it keeps track of the schedulers, clients, and connections to the engines. It handles all the connections of the clients and engines and also the entire traffic. It also maintains a persistent database of all the requests and results that will be utilized in subsequent phases of the applications. The hub provides the functionality to query the state of the cluster, and hides the actual details of a number of connections among the clients and engines.

- **Schedulers**: The Python commands, submitted to engines for processing, are directed through the schedulers. The schedulers also solve the problem of engine blockages while executing user code. Schedulers manage to keep this problem hidden from users and provide fully asynchronous access to the collection of IPython engines.

IPython view and interfaces

The controller provides two interfaces to interact with the engines. The first interface is the `Direct` interface, wherein the engines are directly addressed for task assignment. The second interface is the `LoadBalanced` interface, wherein the proper assigning of tasks to the engines is left to the scheduler. IPython's flexible design enables us to extend the view for implementation of more sophisticated interface schemes.

For a different way of connection to the controller, there is a `view` object. The following are two models for interaction with the machines through the controller:

- The `DirectView` class, which supports direct addressing. It allows command execution on all the engines.

- The `LoadBalancedView` class takes care of task farming on behalf of users in a load balancing method. It allows command execution on any one engine, and the engine to be used for the execution is decided by the scheduler.

The IPython client

The client is an object that is used to connect to the cluster. During the creation of the client object, the user can choose any of the two views discussed previously. Once the client is created, it will be alive as long as the job runs. When the timeout period completes or the user calls the `kill` function, it gets destroyed.

Example of performing parallel computing

The following program is a simple example of performing parallel computing using IPython. It calculates the power on the cluster in a single engine or in parallel in all the engines. Before executing this program, you are advised to check whether the `zmq` package is installed or not, as it is required.

To run these programs in IPython, first start the IPython cluster using the `ipcluster start --n=4 --profile=testprofile` command. It will create the `ipcontroller-engine.json` and `ipcontroller-client.json` files in the `<userhome>/.ipython/profile_testprofile/security` directory. These files will be searched through when we create the client by passing `profile='testprofile'`. If we create the client using `parallel.Client()`, then it will search for JSON files in the `profile_default` folder.

First, the program defines a function to calculate power, and then it creates a client using a test profile. To call a Python function in an engine, we can use the `apply` method of the client or view. The Python `map` function performs serial computation on the sequence. There are `map` function in both `DirectView` and `LoadBalancedView` that performs parallel computation on a sequence. We can also perform these calls in blocking or nonblocking mode. To set blocking mode, we can set the `block` attribute of the client or view to `true`; by default, it is `false`:

```
from IPython import parallel
def pow(a, b):
  return a ** b
clients = parallel.Client(profile='testprofile')
print clients.ids
clients.block = True
clients[0].apply(pow, 2, 4)
clients[:].apply(pow, 2, 4)
map(pow, [2, 3, 4, 5], [2, 3, 4, 5])
view = clients.load_balanced_view()
view.map(pow, [2, 3, 4, 5], [2, 3, 4, 5])
```

A parallel decorator

There is a parallel decorator in `DirectView` that creates the `parallel` function. This function operates on the sequence and breaks up the element-wise operations. Later, it distributes them for parallel computation, and finally, it reconstructs the result. The decorator of `LoadBalancedView` turns the Python function into the `parallel` function:

```
from IPython import parallel
clients = parallel.Client(profile='testprofile')
lbview = clients.load_balanced_view()
lbview.block = True
serial_computation = map(lambda i:i**5, range(26))
parallel_computation = lbview.map(lambda i: i**5, range(26))
@lbview.parallel()
def func_turned_as_parallel(x):
    return x**8
func_turned_as_parallel.map(range(26))
```

IPython's magic functions

IPython has a number of magic functions that a user can call as commands. There are two types of magic commands in IPython, namely line magic and cell magic. Line magic functions are prefixed with % and perform their functionality just like an operating system command. Whereas, cell magic functions are prefixed with %%, and they take the remaining line and the lines after it as different arguments.

These magic functions become available when the user creates a client. The description of line magic functions is as follows:

- %px: This can execute a single Python command on the selected engines. The user can select engines by setting the target attribute of the view instance.

- %pxconfig: Even if we don't have any active view, we can specify --targets, --block, and -noblock using the pxconfig magic function.

- %autopx: This works as a toggling switch for parallel and nonparallel mode. At the first call, it will switch the console to a mode in which all the typed commands/function calls will be executed in parallel mode until the user calls %autopx again.

- %pxresult: In nonblocking mode, %px doesn't return the result. We can see the result of the latest command using the pxresult magic command.

In the cell magic mode, px (%%px) magic accepts the --targets option to specify the target engines to be used, and --block or --noblock to specify the blocking or nonblocking execution mode. This is especially useful in the case where we don't have the view instance. It also has an argument, --group-output, that can manage the presentation of the output of multiple engines.

The following program illustrates the use of px and pxresult as line magic and cell magic. It also covers the autopx and pxconfig line magic and creates specific suffixes for these kinds of magic. The second and third lines of the program perform an import on the IPython session and all the engines. All the imports inside the block created after the second line will also be performed on the engines:

```
from IPython import parallel
drctview = clients[:]
with drctview.sync_imports():
    import numpy
clients = parallel.Client(profile='testprofile')
drctview.activate()
drctview.block=True
%px dummymatrix = numpy.random.rand(4,4)
```

```
%px eigenvalue = numpy.linalg.eigvals(dummymatrix)
drctview['eigenvalue']

%pxconfig --noblock
%autopx
maximum_egnvals = []
for idx in range(50):
    arr = numpy.random.rand(10,10)
    egnvals = numpy.linalg.eigvals(arr)
    maximum_egnvals.append(egnvals[0].real)
%autopx
%pxconfig --block
%px answer= "The average maximum eigenvalue is: %f"%(sum(maximum_
egnvals)/len(maximum_egnvals))
dv['answer']

%%px --block --group-outputs=engine
import numpy as np
arr = np.random.random (4,4)
egnvals = numpy.linalg.eigvals(arr)
print egnvals
egnvals.max()
egnvals.min()

odd_view = clients[1::2]
odd_view.activate("_odd")
%px print "Test Message"
odd_view.block = True
%px print "Test Message"
clients.activate()
%px print "Test Message"
%px_odd print "Test Message"
```

Activating specific views

These magic functions are, by default, associated with a DirectView object. The user is allowed to change the DirectView object by calling the activate() method on any specific view. While activating a view, we can mention a new suffix, such as what is defined in odd_view.activate("_odd"). For this view, we now have a new set of magic functions along with the original magic functions, such as %px_odd, which is used in the last line of the preceding program.

Engines and QtConsole

The px magic function allows users to connect QtConsole to engines for debugging purposes. The following program fragment demonstrates connecting QtConsole to engines by binding the engine's kernel to listen for a connection:

```
%px from IPython.parallel import bind_kernel; bind_kernel()
%px %qtconsole
%px %connect_info
```

Advanced features of IPython

In subsequent subsections, we will have discussions of the various advanced features of IPython.

Fault-tolerant execution

The IPython task interface prepares the engines as fault-tolerant and dynamic-load-balanced cluster systems. In the task interface, the user does not have access to the engine. Instead, task allocation completely depends on the scheduler, and this makes the design of the interface simple, flexible, and powerful.

If a task fails in IPython, for any reason, then the task will be requeued and its execution will be attempted again. A user can configure the system to take a predefined number of retries if there is a failure, and they can also resubmit the task.

If required, users can explicitly resubmit any task. Alternatively, they can set a flag to retry the task for a predefined number of times — by setting a flag of the view or scheduler.

If the user is sure that the cause of the error is not a bug or problem in the code, then they can set the retries flag to any integer value from 1 to the total number of engines.

The reason for the maximum limit being equal to the number of engines is that the task will not be resubmitted to the engine on which it has failed.

There are two options for setting the flag value for the number of resubmissions. One is setting all subsequent tasks after setting the value using the LoadBalancedView (consider the object name to be lbvw) object, as follows:

```
lbvw.retries = 4
```

The other is setting the value using `with ...temp_flags` for a single block, like this:

```
with lbvw.temp_flags(retries=4):
    lbview.apply(task_tobe_retried)
```

Dynamic load balancing

The scheduler can also be configured to perform scheduling on the basis of various scheduling policies. IPython supports a number of schemes to allocate the task to the machine in the case of a load balancing request. It is also very easy to integrate custom schemes. There are two ways of selecting a scheme. One is to set the `taskSchedulerscheme_name` attribute of the controller's `config` object. The second option is to select the scheme by passing the scheme argument to `ipcontroller`, as follows:

```
ipcontroller --scheme=<schemename>
```

Here is an example:

```
ipcontroller --scheme=lru
```

The `<schemename>` function can be any of the following:

- `lru`: **Least Recently Used (LRU)** is a scheme that assigns the task to the least recently used engine.

- `plainrandom`: In this scheme, the scheduler randomly picks an engine to run the task.

- `twobin`: This scheme uses NumPy functions to assign the task. It is the combination of `plainrandom` and `lru`, as it randomly picks two engines and selects the least recently used out of the two.

- `leastload`: This scheme is the default scheme of the scheduler. It assigns the task to the least loaded engine (that is, the engine that has the least number of remaining tasks).

- `weighted`: This scheme is a variant of the `twobin` scheme, as it randomly picks two engines and assigns the load or number of outstanding tasks as an inverse of the weight. It assigns the task to the engine that has a comparatively lesser load.

Pushing and pulling objects between clients and engines

Besides calling functions and executing code on engines, IPython allows users to move Python objects among the IPython client and engines. The push method pushes the objects from clients to the engines, and the pull method can be used to pull back any object from the engines to the clients. In nonblocking mode, push and pull returns the AsyncResult objects. To display the result in nonblocking mode, we can pull the objects as follows: rslt = drctview.pull(('a','b','c')). We can call rslt.get() to display the values in the pulled object. In several cases, it is highly useful to partition the input data sequence and push different partitions to different engines. This partitioning is implemented as scatter and gather functions, similar to MPI. The scatter operation is used to push the partitioned sequence from the client (IPython session) to the engines, and the gather operation is used to fetch the partitions back to the client from the engines.

All of this functionality is demonstrated in the following program. At the end, a parallel dot product of two matrices is implemented using scatter and gather:

```python
import numpy as np
from IPython import parallel
clients = parallel.Client(profile='testprofile')
drctview = clients[:]
drctview.block = True
drctview.push(dict(a=1.03234,b=3453))
drctview.pull('a')
drctview.pull('b', targets=0)
drctview.pull(('a','b'))
drctview.push(dict(c='speed'))
drctview.pull(('a','b','c'))
drctview.block = False
rslt = drctview.pull(('a','b','c'))
rslt.get()

drctview.scatter('a',range(16))
drctview['a']
drctview.gather('a')

def paralleldot(vw, mat1, mat2):
    vw['mat2'] = mat2
    vw.scatter('mat1', mat1)
    vw.execute('mat3=mat1.dot(mat2)')
    return vw.gather('mat3', block=True)
a = np.matrix('1 2 3; 4 5 6; 7 8 9')
b = np.matrix('4 5 6; 7 8 9; 10 11 12')
paralleldot(drctview, a,b)
```

The following program demonstrates the methods that enable the pushing of an object from clients to engines and pulling of the results back from engines to clients. It performs the dot product of two matrices on all the engines and, in the end, collects the results. It also verifies that all the results are the same using the `allclose()` methods, and then returns `True` if the objects are the same. In the `execute` command in the following program, the `print mat3` statement is added with the purpose of displaying the output of the standard output devices of all the engines using the `display_outputs()` method:

```
import numpy as np
from IPython.parallel import Client
ndim = 5
mat1 = np.random.randn(ndim, ndim)
mat2 = np.random.randn(ndim, ndim)
mat3 = np.dot(mat1,mat2)
clnt = Client(profile='testprofile')
clnt.ids
dvw = clnt[:]
dvw.execute('import numpy as np', block=True)
dvw.push(dict(a=mat1, b=mat2), block=True)
rslt = dvw.execute('mat3 = np.dot(a,b); print mat3', block=True)
rslt.display_outputs()
dot_product = dvw.pull('mat3', block=True)
print dot_product
np.allclose(mat3, dot_product[0])
np.allclose(dot_product[0], dot_product[1])
np.allclose(dot_product[1], dot_product[2])
np.allclose(dot_product[2], dot_product[3])
```

Database support for storing the requests and results

The IPython hub stores information about the requests and the results of processing tasks for later use. Its default database is SQLite, and at present, it supports MongoDB and an in-memory database called `DictDB`. The users have to configure the database to be used for their profile. In the active profile folder, there is a file called `ipcontroller_config.py`. This file will be created when we start `ipcluster`. This file has a `c.HubFactory.db_class` entry; users are supposed to set it to the database of their choice, as follows:

```
#dict-based in-memory database named as dictdb
c.HubFactory.db_class = 'IPython.parallel.controller.dictdb.DictDB'
# For MongoDB:
```

```
c.HubFactory.db_class = 'IPython.parallel.controller.mongodb.MongoDB'
# For SQLite:
c.HubFactory.db_class = 'IPython.parallel.controller.sqlitedb.
SQLiteDB'
```

The default value of this attribute is NoDB, which signifies that no database will be used. To get the result of any executed task, the user can call the get_result function on the client object. The client object has a better method called db_query() for getting more insights into the task's results. This method is designed in the MongoDB query style. It takes a dictionary query object with keys from the list of the TaskRecord keys with the exact value or MongoDB queries. These arguments follow the {'operator' : 'argument(s)'} syntax. It also has an optional argument named keys. This argument is used to specify the keys to be retrieved. It returns a list of TaskRecord dict. By default, it retrieves all keys except the buffers for the request and the result. The msg_id key will always be included in the response, similar to MongoDB. Various TaskRecord keys are explained in the following list:

- msg_id: This value is the uuid (bytes) type. It represents the message ID.
- header: This value is the dict type, and it holds the request header.
- content: This value is the dict type, and it holds the request content that will be generally empty.
- buffers: This value is the list (bytes) type, and it will be a buffer containing serialized request objects.
- Submitted: This value is the datetime type, and it holds the submission timestamp.
- client_uuid: This value is the uuid (universally unique identifier in bytes).
- engine_uuid: This value is the uuid (bytes) type that holds the identification of the engine's socket.
- started: This value is the datetime type, and it holds the time when the task execution was started on an engine.
- completed: This value is the datetime type, and it holds the time when the task execution was finished on an engine.
- resubmitted: This value is the datetime type, and it holds the time of resubmission of the task, if it is applicable.
- result_header: This value is the dict type, and it holds the header of the result.
- result_content: This value is the dict type, and it holds the content of the result.

- **result_buffers**: This value is the list(bytes) type and it will be a buffer containing serialized result objects.

- **queue**: This value is the bytes type, and it represents the name of the queue for the task.

- **stdout**: This is a stream of **standard output (stdout)** data.

- **stderr**: This is a stream of **standard error (stderr)** data.

The following program demonstrates the concept of the db_query() and get_result() methods for accessing the result information:

```
from IPython import parallel
from datetime import datetime, timedelta
clients = parallel.Client(profile='testprofile')
incomplete_task = clients.db_query({'complete' : None}, keys=['msg_
id', 'started'])
one_hourago = datetime.now() - timedelta(1./24)
tasks_started_hourago = clients.db_query({'started' : {'$gte' : one_
hourago },'client_uuid' : clients.session.session})
tasks_started_hourago_other_client = clients.db_query({'started'
: {'$le' : hourago }, 'client_uuid' : {'$ne' : clients.session.
session}})
uuids_of_3_n_4 = map(clients._engines.get, (3,4))
headers_of_3_n_4 = clients.db_query({'engine_uuid' : {'$in' : uuids_
of_3_n_4 }}, keys='result_header')
```

The following relational operators are supported in db_query as MongoDB:

- '$in': This represents an *in* operation on the list/sequence

- '$nin': This represents a *not in* operation on the list/sequence

- '$eq': This is used to represent *equal to* (==)

- '$ne': This is used to represent *not equal to* (!=)

- '$gt': This is used to represent *greater than* (>)

- '$gte': This is used to represent *greater than or equal to* (>=)

- '$lt': This is used to represent *less than* (<)

- '$lte': This is used to represent *less than or equal to* (<=)

Using MPI in IPython

Generally, parallel algorithm running on multiple engines requires movement of data among the engines. We have already covered IPython's built-in way of performing this data movement. However, this is a slow operation as it is not a direct transfer between the clients and the engines. The data has to be transferred through the controller. A better way of achieving good performance is by using a **message passing interface** (**MPI**). IPython's parallel computing has excellent support for integration with MPI. To use MPI with IPython parallel computing, we need to install an MPI implementation such as OpenMPI or MPICH2/MPICH and the mpi4py python package. After the installation, test whether the system is able to execute the mpiexec or mpirun command.

After testing the installation and before actually running the MPI programs, the user is required to create a profile for MPI execution using the following command:

```
ipython profile create --parallel --profile=mpi
```

After profile creation, add the following line to ipcluster_config.py in the profile_mpi folder:

```
c.IPClusterEngines.engine_launcher_class = 'MPIEngineSetLauncher'
```

Now, the system is ready to execute MPI-based programs on IPython. The user can start the cluster using the following command:

```
ipcluster start -n 4 --profile=mpi
```

The preceding command starts the IPython controller and uses mpiexec to start four engines.

The following program defines a function that calculates the sum of a distributed array. Save the file with the name as parallelsum.py, as this name will be used in the next program, which actually calls this function:

```
from mpi4py import MPI
import numpy as np

def parallelsum(arr):
    localsum = np.sum(arr)
    receiveBuffer = np.array(0.0,'d')
    MPI.COMM_WORLD.Allreduce([localsum, MPI.DOUBLE],
        [receiveBuffer, MPI.DOUBLE],
        op=MPI.SUM)
    return receiveBuffer
```

The function defined in the preceding program is now called in order to execute it on multiple engines. This is done to perform a parallel sum of the array:

```
from IPython.parallel import Client
clients = Client(profile='mpi')

drctview = clients[:]
drctview.activate()
#execute the program name passed as argument
drctview.run(parallelsum.py.py')
drctview.scatter('arr',np.arange(20,dtype='float'))
drctview['arr']
# calling of the function
%px sum_of_array = parallelsum(arr)
drctview['sum_of_array']
```

Managing dependencies among tasks

It has strong support for managing dependencies among various tasks. In most scientific and commercial applications, only load balancing schemes are not enough to manage their complexity. These applications require dependencies among multiple tasks. These dependencies describe the specific software, Python module, operating system, or hardware requirements; sequence; timing; and places of execution of a task from the set of the task. IPython supports two kinds of dependencies, namely functional dependency and graph dependency.

Functional dependency

Functional dependency is used to determine whether a particular engine is capable of running a task. This concept is implemented using a special exception, UnmetDependency, from IPython.parallel.error. If a task fails with an UnmetDependency exception, the scheduler doesn't propagate this error to the client. Instead, it handles this error and submits this task to some other engine. The scheduler repeats this process until a suitable engine is found. Moreover, the scheduler doesn't submit the task to an engine twice.

Decorators for functional dependency

Although the user is allowed to manually raise an UnmetDependency exception, IPython has provided two decorators to manage this dependency issue. There are two decorators and a class used for functional dependencies:

- @require: This decorator manages the dependency of a task that requires that a particular Python module, local function, or local object be available on an engine when the decorated function is called. Functions will be pushed to the engine with their names, and objects can be passed using the arg keyword. We can pass the names of all the Python modules required to execute this task. Using this decorator, a user can define a function to be executed on only those engines where the module names passed to this decorator are available and importable.

  For example, the function defined in the following code fragment depends on the NumPy and pandas modules as it is using randn from NumPy and Series from pandas. If, for some task, we call this function, then it will be executed on the machine where these two modules are importable. When this function is called, NumPy and pandas will be imported.

```
from IPython.parallel import depend, require
# the following function uses randn and Series
@require('pandas', 'numpy')
def func_uses_functions_from_numpy_pandas():
  return performactivity()
```

- @depend: This decorator allows users to define a function that has a dependency on some other function. It determines whether the dependency is met or not. Before starting the task, the dependency function will be called, and if this function returns true, then the actual processing of the task will be started. Moreover, if this dependency function returns false, then the dependency is considered to be unmet and the task is propagated to some other engine.

  For example, the following code fragment first creates a dependency function that verifies that the operating system of the engine matches the given operating system. This is determined because the user wishes to write two different functions to perform the specific activity on the Linux and Windows operating systems:

```
from IPython.parallel import depend, require
def find_operating_system(plat):
    import sys
    return sys.platform.startswith(plat)
@depend(find_operating_system, 'linux')
def linux_specific_task():
    perform_activity_on_linux()
@depend(platform_specific, 'win')
def linux_specific_windows():
    perform_activity_on_windows()
```

Graph dependency

There is another important class of dependencies, where tasks are dependent on each other in such a manner that a task must be executed after some or all specific tasks have been executed successfully. Another dependency may be as follows: a task must be executed on a destination where a specific set of dependencies has been met. Generally, the user requires an option to specify the time and location to run a given task as a function of time, location, and result of some other task. There is a separate class named as `Dependency` to manage graph dependency and `Dependency` is the subclass of class `Set`. It contains set of message IDs corresponding to the tasks and it also has some attributes. These attributes help in checking whether the specified dependencies have been met or not:

- `any|all`: These are the attributes to specify that if any of the specified dependencies is completed or has been met. This will be specified by setting all attributes of dependence that default to `True`.

- `success`: This attribute defaults to `True` and is used to specify that a dependency is considered to be met if the specified task is successful.

- `failure`: This attribute defaults to `False` and is used to specify that a dependency is considered to be met if the specified task has failed.

- `after`: This attribute is used to specify that the dependent task should be executed after the execution of the specified task.

- `follow`: The `follow` attribute specifies that the dependent task should be executed on the same destination as one of the dependency tasks.

- `timeout`: This attribute is used to specify the duration for which the scheduler must wait for the dependencies to be met. It defaults to `0` to indicate that the dependent task will wait forever. After the timeout period, the dependent task fails with the `DependencyTimeout` exception.

There are some tasks that work as cleanup tasks. They are supposed to run only when the specified task has failed. The user should use `failure=True, success=False` for such tasks. For some dependent tasks, it is required that the dependency task should be completed successfully. In such situations, the user must set `success=True` and `failure=False`. There are certain situations where the user wants the dependent task to be executed independently of the success or failure of the dependency tasks. In such situations, the user must use `success=failure=True`.

Impossible dependencies

There may some dependencies that are impossible to meet. If this possibility is not handled by the scheduler, then the scheduler may wait forever for the dependency to be met. To cope up with this situation, the scheduler analyzes the graph dependency to estimate the possibility that the dependency can be met. If the scheduler is able to identify that the dependency of a certain task cannot be met, then this task will fail with an ImpossibleDependency error. The following code fragment demonstrates the use of the graph dependency among tasks:

```
from IPython.parallel import *
clients = ipp.Client(profile='testprofile')
lbview = clients.load_balanced_view()

task_fail = lbview.apply_async(lambda : 1/0)
task_success = lbview.apply_async(lambda : 'success')
clients.wait()
print("Fail task executed on %i" % task_fail.engine_id)
print("Success task executed on %i" % task_success.engine_id)

with lbview.temp_flags(after=task_success):
    print(lbview.apply_sync(lambda : 'Perfect'))

with lbview.temp_flags(follow=pl.Dependency([task_fail, task_success],
failure=True)):
    lbview.apply_sync(lambda : "impossible")

with lbview.temp_flags(after=Dependency([task_fail, task_success],
failure=True, success=False)):
    lbview.apply_sync(lambda : "impossible")

def execute_print_engine(**flags):
    for idx in range(4):
        with lbview.temp_flags(**flags):
            task = lbview.apply_async(lambda : 'Perfect')
            task.get()
            print("Task Executed on %i" % task.engine_id)

execute_print_engine(follow=Dependency([task_fail, task_success],
all=False))
execute_print_engine(after=Dependency([task_fail, task_success],
all=False))
execute_print_engine(follow=Dependency([task_fail, task_success],
all=False, failure=True, success=False))
execute_print_engine(follow=Dependency([task_fail, task_success],
all=False, failure=True))
```

The DAG dependency and the NetworkX library

Generally, it is better to represent parallel workflows in terms of **Directed Acyclic Graph (DAG)**. Python has a popular library called NetworkX for using graphs. The graph is a collection of nodes and directed edges. The edges connect various nodes; each edge has an associated direction. We can use this concept to represent dependencies. For example, edge(task1, task2) from task1 to task2 denotes that task2 depends on task1. Similarly, edge(task2, task1) denotes that task1 depends on task2. This graph must not contain any cycle in it, which is why it is called an acyclic graph.

Now consider the six-node DAG depicted in the following figure. It indicates that **Task0** doesn't depend on any task, and so it can be started immediately. Whereas, **Task1** and **Task2** depend on **Task0**, so they will start after **Task0** finishes. Then, **Task3** depends on both **Task1** and **Task2**, so it will be executed after the end of **Task1** and **Task2**. Similarly, **Task4** and **Task5** will be executed after the end of **Task3**. **Task6** depends on **Task4** only. Hence, it will be started after **Task4** finishes its execution.

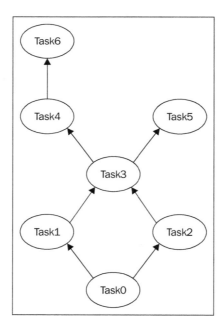

Here is a source code fragment that represents the DAG depicted in the preceding figure. In the code, the tasks are represented by their number; that is, **Task0** is represented by 0, **Task1** is represented by 1, and so on:

```
import networkx as ntwrkx
import matplotlib.pyplot as plt

demoDAG = ntwrkx.DiGraph()
map(demoDAG.add_node, range(6))
demoDAG.add_edge(0,1)
demoDAG.add_edge(0,2)
demoDAG.add_edge(1,3)
demoDAG.add_edge(2,3)
demoDAG.add_edge(3,4)
demoDAG.add_edge(3,5)
demoDAG.add_edge(4,6)
pos = { 0 : (0,0), 1 : (-1,1), 2 : (1,1), 3 : (0,2), 4 : (-1,3), 5 :
(1, 3), 6 : (-1, 4)}
labels={}
labels[0]=r'$0$'
labels[1]=r'$1$'
labels[2]=r'$2$'
labels[3]=r'$3$'
labels[4]=r'$4$'
labels[5]=r'$5$'
labels[6]=r'$6$'

ntwrkx.draw(demoDAG, pos, edge_color='r')
ntwrkx.draw_networkx_labels(demoDAG, pos, labels, font_size=16)
plt.show()
```

The following program creates the same diagram with colored edges and the vertex with the labels of vertices:

```
import networkx as ntwrkx
import matplotlib.pyplot as plt

demoDAG = ntwrkx.DiGraph()
map(demoDAG.add_node, range(6))

pos = { 0 : (0,0), 1 : (-1,1), 2 : (1,1), 3 : (0,2), 4 : (-1,3), 5 :
(1, 3), 6 : (-1, 4)}
```

```
ntwrkx.draw(demoDAG, pos)
ntwrkx.draw_networkx_edges(demoDAG,pos,
                    edgelist=[(0,1),(0,2),(1,3),(2, 3),(3,
4)],edge_color='r')
ntwrkx.draw_networkx_edges(demoDAG,pos,
                    edgelist=[(3,5),(4,6)],edge_color='b')

ntwrkx.draw_networkx_nodes(demoDAG,pos,
                    nodelist=[0,1,2,3,4],
                    node_color='r',
                    node_size=500,
                 alpha=0.8)
ntwrkx.draw_networkx_nodes(G,pos,
                    nodelist=[5,6],
                    node_color='b',
                    node_size=500,
                 alpha=0.8)

labels={}
labels[0]=r'$0$'
labels[1]=r'$1$'
labels[2]=r'$2$'
labels[3]=r'$3$'
labels[4]=r'$4$'
labels[5]=r'$5$'
labels[6]=r'$6$'

ntwrkx.draw_networkx_labels(demoDAG, pos, labels, font_size=16)
plt.show()
```

Using IPython on an Amazon EC2 cluster with StarCluster

StarCluster is designed and developed to simplify the process of using a cluster of virtual machines on Amazon **Elastic Compute Cloud** (**EC2**). It is an open source toolkit for cluster computing on Amazon EC2. Besides performing automatic cluster configuration, StarCluster also provides **Amazon Machine Images** (**AMIs**) customized to support toolkits and libraries for scientific computing and software development. These AMIs consist of ATLAS, IPython, NumPy, OpenMPI, SciPy, and others. The user can retrieve the list of available AMIs using the following command on a machine that has StarCluster installed:

```
starcluster listpublic
```

StarCluster has a very simple and intuitive interface for elastic management of computing cluster and storage management. After the installation, the user must update its default config file to update the details of the Amazon EC2 account, including the address, region, credentials, and public/private key pair.

After the installation and configuration, the user can control the Amazon EC2 cluster from a local IPython installation using the following command:

```
starcluster shell --ipcluster=clusterName
```

If there is any error in the configuration, it will be displayed by the preceding command. If the configuration is correct, then this command starts the development shell of StarCluster and configures a parallel session for the remote cluster on Amazon EC2. StarCluster automatically creates a parallel client by the name of `ipclient` and views for the entire cluster by the name of `ipview`. The user can use these variables (`ipclient` and `ipview`) to run the parallel task on the Amazon EC2 cluster. The following code fragment displays the engine IDs of the cluster using `ipclient` and runs a small parallel task using `ipview`:

```
ipclient.ids
result = ipview.map_async(lambda i: i**5, range(26))
print result.get()
```

The user is also allowed to use IPython parallel scripts with StarCluster. If the user wants to run the IPython parallel scripts on the remote Amazon EC2 cluster from the local IPython session, then they are supposed to use some configuration details during the creation of parallel clients, as follows:

```
from IPython.parallel import Client
remoteclients = Client('<userhome>/.starcluster/
ipcluster/<clustername>-<yourregion>.json',   sshkey='/path/to/cluster/
keypair.rsa')
```

Specifically, suppose the name of a cluster is `packtcluster`, the region is `us-west-2`, and the `keypair` name is `packtKey`, which is stored in `/home/user/.ssh/packtKey.rsa`. Then, the previous code will be changed to the following:

```
from IPython.parallel import Client
remoteclients = Client('/home/user/.starcluster/ipcluster/
packtcluster-us-west-2.json', sshkey='/home/user/.ssh/packtKey.rsa')
```

After these two lines, all of the remaining code will be executed on the remote cluster on Amazon EC2.

A note on security of IPython

Security issues have been taken care of while designing IPython's architecture. The capability-based client authentication model, along with the SSH-tunneled TCP/IP channels, manages the main potential security issues and allows users to utilize the IPython cluster in open networks.

There is no security provided by ZeroMQ. Hence, SSH tunnels are the main source for establishing a secure connection. The `Client` object fetches information about the establishment of a connection to the controller from the `ipcontroller-client.json` file, and then it creates tunnels using OpenSSH/Paramiko.

It also uses the concept of HMAC digests to sign messages using a shared key that protects the users of shared machines. There is a session object that handles the message protocol. This object verifies the validity of messages using a unique key. By default, this key is a 128-bit pseudo-random number, similar to the number generated by `uuid.uuid4()`. Generally, during parallel computations, the IPython client is used to send Python functions, commands, and data to the IPython engines for execution and processing of data. IPython ensures that only the client is responsible for, and capable of using the capability of the engine. The engine inherits the capability and permission from the user that started the engine.

To prevent unauthorized access, authentication- and key-related information is encoded in the JSON file that is used by clients to get access to the IPython controller. A user can grant access to authorized persons by limiting the key's distribution.

Well-known parallel programming styles

Owing to the evolution of computer hardware and software, parallel programs can be designed, developed, and implemented using several styles. We can implement a program using the concurrent, parallel, or distributed manner. Often, one of the previously mentioned techniques is used to implement programs for efficient execution and improved performance. The subsequent subsections discuss these models and the common issues associated with them.

Issues in parallel programming

All of these models depend on the basic concepts of execution of different parts of the program on separate computing elements (CPU and computing node). Generally, these models divide the program into multiple workers, and each worker starts its execution on a different computing element. In spite of the performance benefits, this type of execution of programs — using multiple workers — brings forth multiple complications in communication. This issue is called **Inter-process Communication (IPC)**.

There are several classic IPC problems that require the attention of the developer, namely deadlock, starvation, and race condition:

- **Deadlock**: A deadlock is a condition wherein two or more workers are in an infinite waiting state to acquire the resources occupied by the other waiting worker. There are four necessary and sufficient conditions for this, namely mutual exclusion, hold and wait, no pre-emption, and circular wait. If these conditions occur during the execution of any program, then the program gets blocked and cannot continue the execution:

    ° Mutual exclusion means that the resources are non-sharable

    ° By hold and wait, we mean that each of the workers under the deadlock is holding some resources and requesting for some additional resources

    ° No pre-emption means that the resources allocated to a worker cannot be pre-empted and allocated to another worker

    ° Finally, by circular wait, we mean that the workers under the deadlock form a chain, or circular list, in which each worker is waiting for the resources held by the next worker on the list

- **Starvation**: This condition arises when multiple workers compete for a single resource. In such a situation, each worker is assigned a priority for resource allocation. Sometimes, this priority assignment becomes unfair, and the execution of some of the workers is delayed for a long time. Consider a situation where there are two types of workers competing for a resource: the workers with high priority and workers with low priority. If there is a continuous arrival of high-priority workers, then there might be a situation where some low-priority workers suffer a long (infinite) wait to get the resource released by high-priority workers.

- **Race condition**: This issue arises when there are multiple workers performing both read and write operations on common data and there is a lack of synchronization among the operations performed by them. As an example, suppose two workers read a common piece of data from a database, modify its value, and then write this data back to the database. If these operations are not performed in a well-synchronized sequence, they will leave the database in an inconsistent state.

There are certain techniques that can be used to avoid these issues, although their detailed discussion is beyond the scope of this book. Let's discuss the types of parallel computing in subsequent subsections.

Parallel programming

In the parallel style of program development, the program is divided into multiple workers that execute on separate CPUs without competing for one CPU, as presented in the following figure. These CPUs may be individual processors of multicore computers, or they may be on separate computers and use a technique such as a **Message Passing Interface (MPI)** for communication.

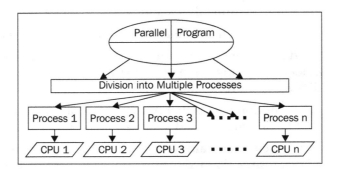

Concurrent programming

In concurrent programming, multiple workers of the user's program get executed on a single CPU or fewer CPUs than the number of workers (as depicted in the next figure). These workers compete for CPUs under the control of a CPU scheduler. The CPU scheduler uses multiple schemes to allocate the workers to the CPUs. The CPU scheduler schemes are used to create a ranking of workers, and the workers get execution in the order of this rank.

These workers may be implemented using multiple processes or multiple threads. Both processes and threads concurrently perform some part of the execution of the main program. The main difference between threads and processes is that a thread consumes less memory than a process. Hence, threads are called **lightweight**.

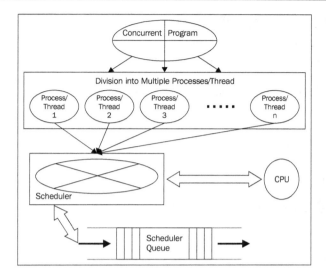

Distributed programming

In distributed programming, the workers of the program execute on different computers connected across a network. There are different frameworks for the execution of these programs. The network also uses different topologies, and in some cases, both the scheme data and the process may be distributed. This model of parallel computing is becoming more popular with time, as it offers several advantages, including low cost, fault tolerance, high scalability, and others. In distributed programming, each component has independent memory and processing, while in parallel programming, multiple processors/CPUs share common memory.

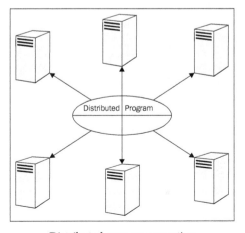

Distributed program execution

Multiprocessing in Python

The multiprocessing module supports the creation and execution of multiple processes that can run independently on multiple CPUs in a multicore environment. It has two important models for supporting multiprocessing; one is based on the `Process` class and the other is based on the `Pool` class.

This program demonstrates multiprocessing using the `Process` class:

```python
import multiprocessing as mpcs
import random
import string


output_queue = mpcs.Queue()

def strings_random(len, output_queue):
  generated_string = ''.join(random.choice(string.ascii_lowercase  +
string.ascii_uppercase + string.digits)
    for i in range(len))
  output_queue.put(generated_string)

procs = [mpcs.Process(target=strings_random, args=(8, output_queue))
for i in range(7)]

for proc in procs:
  proc.start()

for proc in procs:
  proc.join()

results = [output_queue.get() for pro in procs]
print(results)
```

The process-based class returns the results in the sequence of completion of the process. If the user needs to retrieve an ordered result, they have to put in extra efforts, as done in the following code. To obtain an ordered result, another parameter is added to the function and finally to the output. This represents the position or the sequence of the process and, finally, the result is sorted on the parameter. This idea is demonstrated in the following program using the `Process` class and a position parameter in the output:

```python
import multiprocessing as mpcs
import random
import string
```

```
output_queue = mpcs.Queue()

def strings_random(len, position, output_queue):
  generated_string = ''.join(random.choice(string.ascii_lowercase  +
string.ascii_uppercase + string.digits)
    for i in range(len))
  output_queue.put((position, generated_string))

procs = [mpcs.Process(target=strings_random, args=(5, pos, output))
for pos in range(4)]

for proc in procs:
  proc.start()
for proc in procs:
  proc.join()

results = [output_queue.get() for pro in procs]
results.sort()
results = [rslt[1] for rslt in results]
print(results)
```

The `Pool` class provides the `map` and `apply` methods for parallel computation, and it also supports the asynchronous versions of these methods. The `map` and `apply` methods lock the main program until a process has finished, and this concept can be used to produce the sequential output demanded by specific applications.

Multithreading in Python

Python's threading module allows users to create multiple threads of a process to perform concurrent computation. The threads of a process share the same data space with the main process/thread, which enables data sharing and easy communication with each other. Threads are also called lightweight processes, as they require much less memory than processes.

The following program demonstrates the creation and starting of threads:

```
import threading
import time
class demoThread (threading.Thread):
  def __init__(self, threadID, name, ctr):
    threading.Thread.__init__(self)
    self.threadID = threadID
    self.name = name
    self.ctr = ctr
```

```
     def run(self):
       print "Start of The Thread: " + self.name
       print_time(self.name, self.ctr, 8)
       print "Thread about to Exit:" + self.name

  def print_time(threadName, delay, counter):
       while counter:
         time.sleep(delay)
         print "%s: %s" % (threadName, time.ctime(time.time()))
         counter -= 1

thrd1 = demoThread(1, "FirstThread", 4)
thrd2 = demoThread(2, "SecondThread", 5)
thrd1.start()
thrd2.start()
print "Main Thread Exits"
```

This program starts two threads. If you observe the output of the program, you will notice that there is no sequence of thread execution. The `"Main Thread Exits"` string is displayed first, followed by the random sequences of thread names and `"Thread about to Exit: ThreadName"`.

We have options for synchronizing this output so that the order of thread completion can be maintained. The following program first executes the first thread. Then, after its exit, the second thread gets executed. Finally, the main thread exits. The thread sequence is maintained by acquiring a lock, and this lock is released before the exit so that the second thread can be started. The main thread calls the `join` method on all thread objects. This method blocks the main thread for the completion of various other threads:

```
import threading
import time
class demoThread (threading.Thread):
  def __init__(self, threadID, name, ctr):
    threading.Thread.__init__(self)
    self.threadID = threadID
    self.name = name
    self.ctr = ctr
  def run(self):
    print "Start of The Thread: " + self.name
    threadLock.acquire()
    print_time(self.name, self.ctr, 8)
    print "Thread about to Exit:" + self.name
    threadLock.release()
```

```
def print_time(threadName, delay, counter):
    while counter:
        time.sleep(delay)
        print "%s: %s" % (threadName, time.ctime(time.time()))
        counter -= 1

threadLock = threading.Lock()
thrds = []

thrd1 = demoThread(1, "FirstThread", 4)
thrd2 = demoThread(2, "SecondThread", 5)
thrd1.start()
thrd2.start()

thrds.append(thrd1)
thrds.append(thrd2)
for thrd in threads:
    thrd.join()

print "Main Thread Exits"
```

Hadoop-based MapReduce in Python

Hadoop is an open source framework for distributed storage and processing of huge datasets in a computing cluster. The Hadoop system has three main components: MapReduce for processing, the **Hadoop Distributed File System (HDFS)**, and a large-scale database called HBase for storing datasets. HDFS supports storage of very huge dataset files. It distributes the datasets submitted by users on various cluster nodes. It splits the datasets into multiple chunks and keeps bookkeeping information about the chunks and nodes. HBase is a database designed to support large-scale databases and developed on top of HDFS. It is an open source, column-oriented, non-relational, distributed database.

MapReduce is a framework that is designed to perform distributed processing on huge datasets in a computing cluster. Hadoop is an open source implementation of the MapReduce framework. The MapReduce program is composed of two main components: map and reduce. The map function is for performing filtering of the input dataset and writing its sorted output to the filesystem. Later, this output is used by the reduce function to perform summarization, and the final output is again written to the filesystem. The MapReduce framework follows the **Single Program, Multiple Data (SPMD)** model, as it processes the same program on multiple datasets.

In a Hadoop system, the complete functionality is divided into many components. There are two master nodes. One is Job Tracker, which keeps track of the map and reduce processes at the slave nodes, called task tracker nodes. The second master node is namenode, which contains the information about which chunk of the split data files is stored on a particular slave node called data node. To avoid a single point of failure, users may install a secondary namenode. It is recommended to have a number of slave nodes called task tracker nodes/data nodes to execute the actual map and reduce processes. Each slave node behaves as a task tracker node and data node. The performance of a MapReduce application is directly proportional to the number of slave nodes. The Hadoop system also performs automatic failure recovery; if one of the task tracker nodes fails at runtime, then its responsibility will automatically be distributed to the other task tracker, and the processing continues without failure.

The following program demonstrates the development of a Hadoop-based MapReduce program in Python. It processes the common crawl datasets. These datasets contain petabytes of web crawling data collected over a long period. It contains web page data, extracted metadata, and extracted text stored in **Web ARChive (WARC)** format. These datasets are stored on Amazon S3 as part of the Amazon public datasets program. More information about these datasets can be found at http://commoncrawl.org/the-data/get-started/:

```python
import sys
for line in sys.stdin:
  try:
    line = line.strip()
    # split the line into words
    words = line.split()
    # increase counters
    if words[0] == "WARC-Target-URI:" :
      uri = words[1].split("/")
      print '%s\t%s' % (uri[0]+"//"+uri[2], 1)
  except Exception:
    print "There is some Error"
```

The preceding program is the map part, and the following program is the reduce part:

```python
from operator import itemgetter
import sys

current_word = None
current_count = 0
word = None
```

```
for line in sys.stdin:
    line = line.strip()

    word, count = line.split('\t', 1)

    try:
        count = int(count)
    except ValueError:
        continue

    if current_word == word:
        current_count += count
    else:
        if current_word:
            print '%s\t%s' % (current_word, current_count)
        current_count = count
        current_word = word

if current_word == word:
    print '%s\t%s' % (current_word, current_count)
```

Before executing the previous program, the user needs to transfer the input dataset file with the name web-crawl.txt in their HDFS home folder. To execute the program, the user should run the following command:

```
#hadoop jar /usr/local/apache/hadoop2/share/hadoop/tools/lib/hadoop-
streaming-2.6.0.jar -file /mapper.py    -mapper /mapper.py -file /
reducer.py   -reducer /reducer.py -input /sample_crawl_data.txt
-output /output
```

Spark in Python

Spark is a general-purpose cluster computing system. It supports high-level APIs for Java, Python, and Scala. This enables easy writing of parallel tasks. It was proposed and developed in contrast to the two-stage model of Hadoop's disk-based MapReduce, as it follows an in-memory model and provides maximum 100 percent performance improvement for some specific applications. It is highly suitable for implementing machine learning applications/algorithms.

Spark requires cluster management and a distributed storage system. It provides a simple interface for various distributed storages, including Amazon S3, Cassandra, HDFS, and so on. Moreover, it supports standalone—that is, the spark native cluster, Hadoop, YARN, and Apache Mesos for the cluster management.

The **Spark Python API** is named **PySpark**, and it exposes the Spark programming model to Python. We can develop Spark-based applications either in the Python shell that is opened using PySpark, or by using the IPython session. We can also develop the program first and then run it using the `pyspark` command.

Summary

In this chapter, we discussed the concepts of high-performance scientific computing using IPython. We started the discussion from the basic concepts of parallel computing. After the basic concepts, we discussed the detailed architecture of IPython parallel computing. Later, we discussed the development of sample parallel programs, IPython magic functions, and parallel decorators.

We also covered the advanced features of IPython: fault tolerance, dynamic load balancing, managing dependencies among tasks, object movement between clients and engines, IPython database support, using MPI from IPython, and managing the Amazon EC2 cluster using StarCluster from IPython. Then we discussed multiprocessing and multithreading in Python. At the end, we covered the development of distributed applications using Hadoop and Spark in Python.

In the next chapter, we will discuss several real-life case studies of using Python's tools/APIs for scientific computing. We will consider applications from various basic and advanced branches of science.

9
Revisiting Real-life Case Studies

This chapter discusses several case studies of scientific computing applications, APIs/libraries, and tools designed and developed in Python.

In this chapter, we will be discussing some case studies of applying Python in the following areas of science:

- Specialized hardware/software
- Applications for meteorologists
- Designing and modeling
- Applications for High-energy Physics
- Computational chemistry
- Biological science
- Embedded systems

These applications, tools, and libraries cover various social, scientific, and commercial areas, including applications for NGOs, software or hardware for scientific education, and applications for meteorologists. They also cover tools and APIs for conceptual designs of aircraft, applications for earthquake risk assessment, and applications designed for energy efficiency of manufacturing processes. Apart from these, analysis code generators for High-energy Physics, applications for computation chemistry, a blind audio tactile mapping system, tools for air traffic control, embedded systems for energy-efficient lights, maritime designing APIs, and molecular modeling toolkits are also covered.

Scientific computing applications developed in Python

Python is a popular language for developing scientific applications. Specially, it is most suitable for applications that demand low costs and applications that require high performance. In subsequent subsections, we will be covering some applications, tools, and products that use Python in some or other form.

The one Laptop per Child project used Python for their user interface

One Laptop per Child (OLPC) is a project started at the **Massachusetts Institute of Technology (MIT)**. The project has grown with support from people who create software and hardware and by solid community involvement to fulfill the mission of OLPC. The idea behind this project is to develop low-cost educational laptops with innovative hardware and software. The simple and convincing mission of OLPC is to create opportunities of education for poor children by providing them with low-cost, low-powered laptops with software and applications designed for collaborative learning. The primary goal of this mission is the production and distribution of a low-cost and low-power laptop named OLPC XO. This XO is manufactured by Quanta Computer, a Taiwanese company. Unlike other computers, XO uses flash memory instead of a hard drive and a Fedora Linux-based operating system. It also uses mobile ad hoc networking designed on the 802.11s wireless mesh network protocol. The XO laptop is shown in the following image:

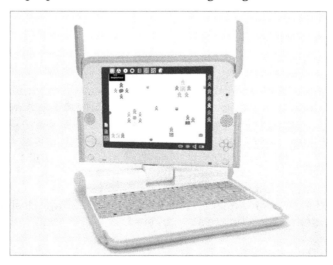

Source: http://images.flatworldknowledge.com/lule/lule-fig13_004.jpg

Sugar is a free and open source desktop environment designed with a noble focus on interactive learning, and it is the interaction interface of XO. Sugar doesn't have the concept of desktop, folders, and windows. Instead of these, it starts with a home view; from this screen, users can select any activity. Applications designed for Sugar are called **activities**. Activities include an application along with sharing and collaboration capabilities. To save the application's state and history, Sugar implements a journal that allows users to restore activities. The journal automatically stores the user's session and provides an interface to retrieve the history by date. Each activity has access to a built-in interface for the journal and other features, such as the clipboard. Sugar's activity runs in full screen mode and lets users use only one program at a time.

Sugar is available for many platforms, as follows:

- **XO laptop**: An XO laptop runs Sugar as the default interface
- **Live CD and Live USB stick**: Sugar is also available on live CDs / USB sticks
- **A package for Linux distributions**: Sugar is available as a package for most Linux platforms—as an alternative desktop
- **An OS image (using virtualization)**: Users can also install Sugar on the Windows or Macintosh operating system using virtualization

The Python language is used to develop Sugar and various activities in Sugar. Developers can use Python to extend Sugar and add new applications/activities. The home screen of Sugar is depicted in the following screenshot:

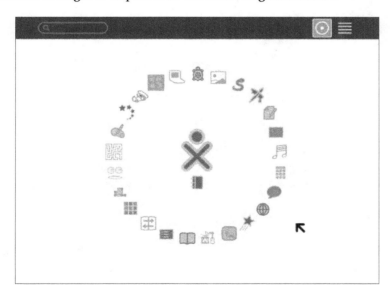

The source of the preceding screenshot is `http://2.bp.blogspot.com/_`
`PPJgknwAe5o/S_8kh3r1qII/AAAAAAAAGk/qmJdLae1pQ8/s1600/2009-SugarLabs-`
`Homeview.png`.

Soon after the launch of XO, it received strong support from the community as it
runs free and is a piece of open source software that allows developers to understand
and improve the software. XO has a high-resolution, easy-to-read screen and
supports a book reader mode and multiple languages. The book reader mode
of the XO laptop is shown in this image:

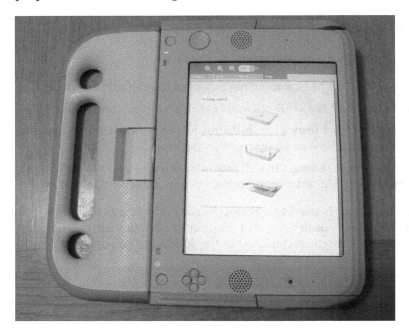

Source: `http://regmedia.co.uk/2008/01/16/ebook.jpg`

ExpEYES – eyes for science

Inter University Accelerator Centre (IUAC) in India has started a project called
Physics with Homemade Equipment and Innovative Experiments (PHOENIX).
The idea behind this project is to improve the quality of scientific education by
experimentation. The major activity of the project is the development of low-cost
laboratory equipment. Another project under this initiative is **Experiments for
Young Engineers and Scientists (ExpEYES)**, designed with the focus on learning by
exploration. ExpEYES is suitable for high school and higher classes. The design is
optimized to meet the main objective, that is, low-cost devices. This device runs on
5-volt USB power.

To use ExpEYES, the user can access it by connecting it to a USB port of a computer and the interface application. It has 32 I/O terminals arranged on both sides for connecting to external signals. The user can control and monitor the voltages at the terminals. To measure other parameters, such as force, pressure, temperature, and so on, the user is supposed to convert them into electrical signals using the appropriate sensor elements. For example, a temperature sensor will give a voltage that indicates the temperature. ExpEYES is shown in the following image:

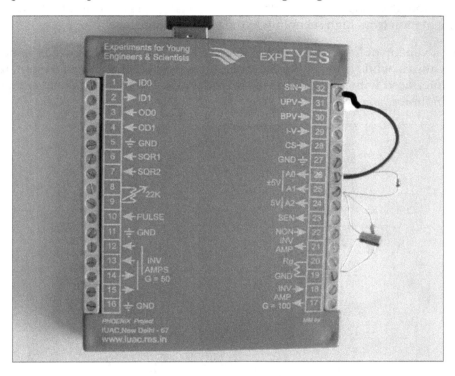

Source: http://expeyes.in/sites/default/files/images/diode-rectifier-photo.jpg

Actual learning requires exploration and performing experimentation. Experiments in physics require controlling and measuring of various parameters, such as acceleration, current, force, pressure, temperature, velocity, voltage, and so on. A number of properties require automated measurements as their values change rapidly (for example, AC mains voltage). These automated measurements require the involvement of computers.

A Python interpreter and a Python module for accessing the serial port are required to run ExpEYES on any computer. The device driver program handles the USB interface; the driver presents the USB port as the RS232 port to application programs. The communication part of ExpEYES is handled by using a library written in the Python language. A GUI program is developed for each of the supported experiments. The user can also develop a Python program to perform new experiments. ExpEYES is available as a live CD and an installation for Linux and Windows. It is an affordable scientific laboratory that is both portable and extensible. It supports a wide range of experimentation from high school up to the post graduate level.

The latest version of ExpEYES is called **ExpEYES Junior**. This version has some added features, while some of the earlier features have been removed from it. It can also be interfaced with Android-based smart devices. The following image depicts ExpEYES Junior:

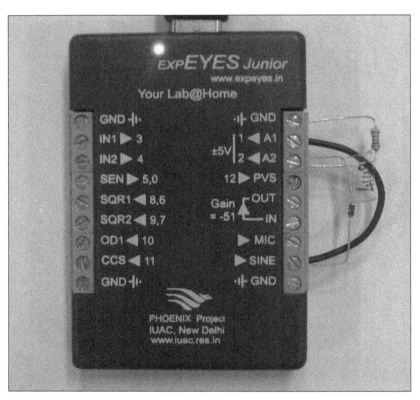

Source: `http://expeyes.in/sites/default/files/Experiments/Photos/half-wave.jpg`

The initial software was written in the C language, but it was soon changed to Python. This change brings with it two major advantages. The first advantage is that support for the development of GUI-based programs is drastically increased. The other advantage is that this has enabled easy development of new experiments because of Python's support for hardware interaction.

A weather prediction application in Python

Generally, meteorologists compare their forecasts with the actual weather for the concerned period. This is performed with the purpose of optimizing and improving the quality of their model, which collects the weather information of the real measurement that collects actual readings. **ForecastWatch** helps meteorologists and others compare, communicate, and understand the accuracy of their weather forecast. It provides essential analyses and unbiased data with the purpose of improving the quality of the forecast. ForecastWatch consistently collects data from various sources of weather forecast and tallies this data with actual observations. It compares each forecast with observations made at more than 850 locations in the United States of America and Canada. In this comparison, high and low temperature, opacity, precipitation, and wind forecasts are verified. ForecastWatch generates a large number of month-wise statistical measures for the accuracy of data and aggregates them by country, state, and specific location.

ForecastWatch is composed of four major components, as follows:

- **Input process for acquiring forecasts**: This is named forecast parser, and it collects the forecasts from the web portal of each forecast provider to be verified. It parses the data and inserts it into the database for comparison with the actual data.

- **Input process for acquiring measured climatological data**: This is termed as the actual parser. It retrieves the actual data provided by the **National Weather Service** from the **National Climatic Data Centre**. This data contains high and low temperatures, precipitation, and significant weather events. The actual parser stores this data in the database and performs the scoring of weather forecasts in comparison with actual data. It stores this information in the database as well.

- **Data aggregation engine**: After collection and scoring, the data is processed by the data aggregation engine so as to arrange it in monthly blocks and yearly blocks separated by location, number of days, and provider.

- **Web application framework**: Initially, the web interface was designed in PHP, and then it was redesigned in Python. Redesigning using Python has simplified the web development and improved its integration with the other three components of the system. A Python-based web application framework called **Quixote** is used to develop pure Python-based web applications.

This is a pure Python-based application, as Python is used to develop all the four components, from an interesting web interface to the time-consuming input/output-bound data collection process and the high-performance aggregation engine. The developers selected Python as it has a number of standard libraries that can be utilized in data collection, parsing, and storing in the database. The multithreading library is used to scale the data collection to a large number of cities in the data collection processes. The aggregation engine has also been developed in Python. It uses a Python database library named **MySQLdb** to execute database queries for the MySQL database that is created by the input processes to store the forecast and climatological data.

An aircraft conceptual designing tool and API in Python

In this section, we are going to discuss a tool and an API developed to support the conceptual designs of aircraft. First, we will be discussing the **VAMPzero** tool. Then, we will discuss the **pyACDT** API.

The **German Aerospace Centre** (abbreviated as **DLR** in German) is the national center for aerospace, energy, and transportation research in Germany. Their focus is to carry out research for aeronautics and space and develop software for related research and development projects. DLR uses Python to develop software tools and APIs.

VAMPzero is a software tool for conceptual designs of aircraft; it has enabled DLR to tackle challenges and provide flexibility and transparency in the aircraft conceptual design process. The requirements in aircraft design change frequently, as they use novel technologies. VAMPzero's flexible process enables users to adopt these changes. VAMPzero is based on well-known handbook methods and is highly extensible. It allows its users to trace back the calculation history and export data in CPACS format. Designing a new system with VAMPzero includes outer geometry, along with engines, structures, systems, and costs. VAMPzero is the first open source tool made for aircraft conceptual design that supports working in a multidisciplinary environment. Programming for VAMPzero is done in Python. It is developed with the objective of having a tool that quickly performs the aircraft design process.

There is a framework developed by the scientists in the **Advanced Aircraft Design Lab** at the **Royal Military College of Canada**, called **Python aircraft conceptual design toolbox (pyACDT)**. pyACDT is a Python-based object-oriented framework developed to perform analyzing, defining, designing, and optimization of aircraft configurations. It consists of several modules for representing each of the major disciplinary analyses needed at the conceptual design stage. This framework represents models of various aircraft components, engine components, characteristics, and disciplinary analyses using the concept of object-oriented programming. The following figure depicts the various modules of pyACDT, corresponding to various disciplines. The design of the framework allows users to easily change constraints, design variables, disciplinary analyses, and objective functions.

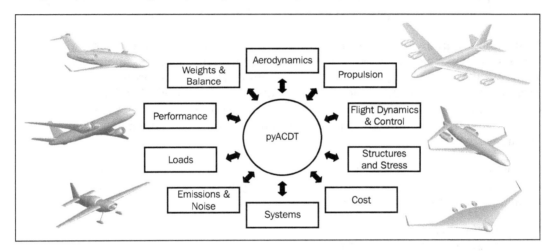

OpenQuake Engine

The Global Earthquake Model is a collaborative effort of several organizations of regional, national, and international levels and a number of individuals to develop uniform and open standards for worldwide calculation and communication of the risk of earthquakes. This foundation is a public-private partnership, in which thousands of people are contributing in all possible ways, including their time and knowledge. This is done to work in various global projects, as a user of the software by using and testing them, reviewing the outcomes of projects and participation in meetings. The activities governed by GEM are highly crucial, as the vulnerability to earthquakes is increasing day by day, and still most of the world is lacking in reliable risk assessment tools and data. Moreover, we lacked global standards to compare various approaches of risk analysis. To get a proper understanding of an earthquake's consequences and behavior, it is considered better to work together across the globe. GEM was created to manage these issues.

The GEM Foundation works the following main areas:

- **Earthquake risk assessment tools**: The primary focus is to design, develop, and enhance high-quality tools for earthquake risk assessment

- **Earthquake risk information**: GEM also works in the direction of collection and generation of methods and guidelines for datasets and models for earthquake risk information

- **Collaborative risk assessment projects**: The GEM foundation works on the development and implementation of collaborative risk assessment projects of various scales

- **Technology transfer and capacity development**: GEM also works for capacity building and knowledge transfer related to earthquake risk assessment

Under the umbrella, scientists are developing best practices, creating common datasets, and developing models for seismic hazard and risk assessment. The GEM foundation is integrating all of these contributions into a web-based OpenQuake toolkit. This toolkit is accessible to the worldwide stakeholders. The OpenQuake engine is developed in the Python language and used by engineers, financial experts, government officials, and scientists to perform earthquake hazard and risk assessment.

OpenQuake is a web-based risk assessment toolkit that offers a single integrated environment to calculate, visualize, and investigate earthquake risks; capture new data; and share the findings for collaborative learning.

The OpenQuake engine uses five main calculators to work in different areas of seismic risk assessment and mitigation. The brief description of these calculators is as follows:

- **Scenario risk calculator**: This calculator is highly useful for raising social awareness toward the risk of earthquakes and for proper emergency planning and management. It is used to calculate the losses and loss statistics for a single earthquake scenario for the given set of assets.

- **Scenario damage assessment calculator**: This calculator is useful for assessment of the seismic vulnerability of various assets under study. It supports the estimation of the damage done to a particular asset from the collection of assets.

- **Probabilistic event-based risk calculator**: This calculator is useful for computing the aggregated expected losses of a collection of assets. It is capable of computing the probability of losses and loss statistics for a given collection of assets using the probabilistic hazard.

- **Classical PSHA-based risk calculator**: The output of this calculator can be utilized to prioritize the risk mitigation efforts for various assets at different locations. This calculator performs the computation of the probability of losses and loss statistics for a single asset. Moreover, these computations for different assets at several locations can be used to perform comparative risk assessment among the assets.

- **Benefit–cost ratio calculator**: This calculator is useful for prioritizing various regions that require strengthening activities and for identifying the seismic design that is economically suitable for a given region. It computes and finds out whether the use of retrofitting or strengthening measures for a specific set of buildings is economically fruitful.

SMS Siemag AG application for energy efficiency

SMS Siemag AG is a market leader in metallurgical plant and rolling mill technology. The company is working with their clients to improve the energy efficiency of their casting plants and their impact on the environment. The company named this operation Eco Mode. In this mode, at a particular time, the devices that are not required for production are automatically shut down or switched to power saving mode. This automatic process is governed by Python-based software. This Python-based application is used to measure and record the power consumption of various consumers. In this way, it performs logging and evaluation of power consumption of the various generator units in different operating modes.

Automated code generator for analysis of High-energy Physics data

The **Large Hadron Collider** (**LHC**) is the world's biggest man-made machine designed to perform particle physics experiments. The main aim of these experiments is to confirm the theories of particle physics and discover new particles. This has started another era of **High-energy Physics** (**HEP**).

This is the biggest scientific experiment ever performed. Here are some facts about it:

- Nearly 100 countries have been involved

- Approximately 500 institutes are collaborating

- Around 10,000 people are working on / benefitting from the experiments

- The cost of the project is about 4 billion Euros

- The length of LHC tunnel is 27 km

This gigantic machine produces large amounts of data of a 10 petabytes/year scale. It is impossible to store and manage so much data at a single site. To cope up with this problem, CERN has developed a grid computing environment in collaboration with almost all of the HEP institutions of the world. This grid is a massive parallel processing system with a network of institution, sharing storage space and processing power. For better performance, the analysis job is transparently executed on the system that has the data to be analyzed.

In the LHC tunnel, two proton beams are circulated. They collide at four experimental points every 25 nanoseconds. As a result of these collisions, many particles are produced. Some of these produced particles are well-known, and it is expected that some others may be new and unknown as yet.

To properly extract the information from this collision data, the physicists have to write codes for each physics signature they are interested in. This code would be efficient only if one signature is to be handled at a time. However, they need to write many such analysis codes to scan all possible new physics signatures. All of these codes are mostly similar and are created by copying the common code; generally, this common code may be *error-prone*, as there are many codes to debug and maintain.

To cope up with this issue, the scientists at CERN have worked on a new idea for the HEP community — to develop a computer-aided software engineering package in Python that takes care of the common code and algorithms and automatically generates analysis code. As this code is automatically generated from user inputs, it proves to be more efficient and less error-prone. It enables physicists to properly take care of the physics part, as the analysis code is automatically generated. This package is named WatchMan. WatchMan is an object-oriented framework that has been developed completely in Python. It allows physicists to focus mainly on their idea, without any worry about analysis code, as it builds complete analysis code from user settings. It is developed using two tools developed in Python at CERN: PyROOT (a Python toolbox for data analysis) and rootcint (the Python and C++ binding system). The process of WatchMan is presented in the following figure:

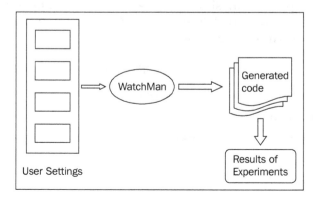

Python for computational chemistry applications

AstraZeneca, a well-known pharmaceutical company, provides effective medicines for cancer, cardiovascular diseases, gastrointestinal and other infections, pain control, and other diseases. Generally, it takes a long time (usually decades) to discover a new drug. The biggest challenge is to identify the possible candidates out of the vast range of molecules that may produce good drugs as early as possible.

There are several techniques for predicting the properties and behavior of molecules. These techniques are developed by computational chemists and used to ensure that a particular molecule will not be toxic to the body and can be stable, will perform the desired activities, and will ultimately get eliminated automatically.

The problem with these techniques is that their results alone are not sufficient, and the chemists have to perform actual experiments. These molecules must be tested in the laboratory to observe their behavior and reaction. Various computational models are used to shortlist the good candidates to be tested to save time.

Before AstraZeneca enhanced the drug identification process, experimental chemists and computational chemists were dependent on each other for drug identification. As experimental chemists did not have much exposure to computational techniques, computation chemists were supposed to help them run the computer predictions, and that is a complex process. This dependency was affecting the progress of both computational and experimental chemists, as computational chemists invested their time to frequently run the routine models instead of developing new techniques for predictions. If there were some technique that enabled experimental chemists to perform computational predictions, then it would improve the process and make the drug prediction process easy and faster.

There was a successful web-based tool designed by Pierre Bruneau using Perl scripts. This tool used a molecular property calculator tool called Drone. It was adopted by AstraZeneca to enhance its backend tool Drone to be more manageable, extensible, and robust. This new backend tool was named **PyDrone**.

Strong explicit error handling and strict type checking enhanced the robustness of Drone when it was implemented in Python (PyDrone). Initially during the testing phase, PyDrone kept throwing exceptions for a number of cases that were silently executed by Drone. The developers found that these exceptions had identified several new error cases that were not handled previously. The new code, with added error handling coding, improved robustness, as many new erroneous cases were handled by the code.

To improve the extensibility of PyDrone, a rule base has been added. This rule base consists of a data cache and a property name for predicting function mapping.

The rule base works like a Python dictionary object. For each requested property, it first looks for the cache. If it is found, then it is used. Otherwise, the associated function is called for computation.

The result will be returned and stored in the cache for future use. For a new prediction, the developer adds the new function to the function table. In this way, PyDrone becomes capable of managing all present and future methods of prediction.

Python for developing a Blind Audio Tactile Mapping System

The **Blind Audio Tactile Mapping System (BATS)** for visually disabled people provides them with access to maps. Before the invention of this system, there were no maps of the ancient world that could be accessible to visually disabled people. The project was started in the University of North Carolina with a small group. Python was selected instead of C++ or Java. This was initially a tough decision, as the team did not have much idea of Python and they were proficient in C++ and Java. Ultimately, this decision was a wise decision, as Python has an extensive collection of libraries and modules that are desirable for the development of such applications.

BATS uses the ArcView data files provided by the Ancient World Mapping Center. ArcView/ArcGIS is a piece of fully functional GIS software used to visualize, manage, create, and analyze geographical data. Initially, they produced two ASCII text files for the surface types and elevation of maps. This information is prepared as a grid of 1024 x 768 that matches the resolution of the display and the touch pad used in the system. This grid information is read and stored in a Python array. The data scales down to fit the BATS model. It is then stored in a compressed file; the program decompresses and loads this data into the appropriate data structure to perform fast startup. There is one-to-one correspondence between the display pixel and the value in the file. This system is highly responsive and promptly takes care of user movements. Various audio/visual effects have been used to represent movements in different areas, such as oceans or land.

BATS is composed of two major components, namely a graphical user interface and a data manager. The data manager enables the user interface to manage the data. The user interface has a touch pad, a number keypad, and a voice synthesizer for helping visually disabled people. User movements on the touch pad are captured through **wxPython**. For user movements, there are mouse motion events in wxPython; these events are used to trigger the query to the surface type and database of various cities. The mouse and key events are handled by wxPython to produce voice feedback. BATS also uses Microsoft's Speech API.

wxPython is cross-platform GUI API that allows Python programmers to develop GUI-based programs with rich user interfaces and event handling.

The data manager stores various values in three numerical arrays and an ODBC connection to the MS Access database. These arrays store the altitude, the land type, and a key value from the database. The key value is used to query the Access database, which retrieves the information on the city that corresponds to the given location.

TAPTools for air traffic control

Developing generalized air traffic control solutions requires extra effort, as the features of each airport are unique in various aspects, such as design, regulatory compliances, and infrastructure. The most significant component of an air traffic control system is its user interface customization.

Frequentis is a frontline solution provider in the field of air traffic management, public safety, and transport. They use Python to develop their TAPTools product family, which works on the tower and airport tools that are part of air traffic control. Air traffic controllers use these to control runway lighting and navigational aid instruments, monitor the navigation instruments, and keep track of the weather conditions.

Designing a fresh user interface for each customer is a tedious and time-consuming task. To manage this problem, Frequentis has developed a tool to design the user interface layout, named PanView. This tool can be used to design and build a user interface that will be executed by the related software, called PanMachine. This software runs on a piece of specially designed hardware called **PowerPanel**. These tools can be used to easily develop a prototype of a layout. Initially, PanView and PanMachine used a scripting language called **Lua**. Lua was used to connect the user interface with the actual functionality of the air traffic control system.

In comparison with Python, there were a number of problems in using Lua. In the event of an error, it provides limited information. It does not have any list data structure, and it is difficult to write larger programs as it has limited standard libraries.

The Finnish Civil Aviation Administration wanted to run the user interface layout not only on PowerPanel but also on web browsers. This project forced the reimplementation of PanMachine in Java to enable the execution in browsers. As Lua cannot be run under Java, Frequentis redesigned it as a Python-based implementation. They selected Python and the Java implementation of Python, called **Jython**. This will enable the user to run the user interface layout on both PowerPanel and PanMachine, which are implemented in Java. For PowerPanel, Python is implemented in C, and Jython is implemented in Java and used for the browser. After this step, the Frequentis developers redesigned the Lua layout in Python. The Python code written for layouts was very short in comparison to the Lua layouts, and thus highly manageable.

Energy-efficient lights with an embedded system

Carmanah Technologies is a leader in the market of solar-powered LED lightings. This company is the manufacturer and supplier of a complete range of lights that can be used for different purposes, including airfield illumination, industrial markers, marine applications, railways, roadways, transits, and others. The company started by producing self-contained and autonomous solar-powered lights for marine navigation. Now, Carmanah's market spans the whole world, specifically places with extreme conditions, such as open oceans, deserts, the far north, and others. These days, electric lighting has become so complex that it should satisfy properties such as autonomous and self-contained light. The usable solar radiation around the light varies with the weather, season, location of the light, orientation of the solar panel, and other attributes. There are some special applications; these lights also support programmable interfaces, provide different outputs according to the inputs, connect to a centralized control station through wireless networking, and are fit for other complex scenarios.

It requires a great combination of electrical, electronic, mechanical, and optical design to develop such lights. Each light is operated using an embedded software program that runs on a microcontroller. These lights are autonomous in the sense that each light maintains itself and performs its functionality as per the modeled requirement.

Generally, embedded systems need components of high reliability, low power consumption, and small size. To fulfil these requirements, special processor chips called microcontrollers have been designed. These microprocessors combine CPU memory and peripherals on a single chip in a very low cost. Besides the embedded functionality written on the microprocessor ROM, there are several functions that require the desktop/laptop system during the development and maintenance periods.

Now, consider an example in which the embedded software is compiled on conventional systems; the object code is then loaded onto the desired microcontroller. Similarly, during maintenance, troubleshooting the deployed device requires additional hardware, such as a laptop, to execute the diagnostic utility. Python offers a number of features to perform the actual and support activities for embedded system development. These features include compactness of Python programs, automated memory management, simple and powerful object-oriented facilities of Python, and so on.

Carmanah uses Python adoption in several key areas of the life cycle of its embedded systems. For example, they use Python programs to control the software building process, stress and unit testing, the device simulator, and other areas.

Scientific computing libraries developed in Python

In Python, a number of libraries are developed for different application domains. These libraries are developed to build commercial applications as well as scientific applications. The following subsections cover some of the selected scientific computing libraries.

A maritime designing API by Tribon

Tribon Solutions works in the area of computer aided design and modeling solutions. Their focus is to improve the overall efficiency of maritime applications. The Tribon software suite supports the complete life cycle of ship building. This requires highly concurrent processes to cope up with the situations. They have developed a central repository and a single source of information for people who work on the design and construction of ships. This model is called **Product Information Model (PIM)**. These people may be designers, material administrators, manufacturing team members, planners, and others involved in the overall process.

In general, ship designs are unique; yet, the focus of their designers is to reduce costs by proving the process through standardization and parameter-driven design. This process is mostly vendor dependent due to different design principles, governmental regulations and standards, and the facilities and infrastructure of the vendor. Tribon handles this problem to allow the vendors to develop on their own. The Tribon technology has created an easy-to-use, platform-independent, extensible, and embeddable API.

Tribon has selected Python because of a number of features, such as the product being embeddable and extensible, no license cost, and platform independence. Tribon's solution doesn't get affected by any updated release of Python. The applications developed by their clients are platform independent, and thus the clients have moved them across platforms without many problems and changes in code. These solutions have improved the design process of some parts from a couple of weeks to a few days, with improved overall quality. This is because the design, computations, and other processes are being automated. They named this product Tribon Vitesse.

Molecular Modeling Toolkit

The **Molecular Modeling Toolkit** (**MMTK**) is a Python library for molecular modeling and simulation with a focus on biomolecular systems. It is an open source library developed using the Python and C languages. Biomolecular simulations take a longer time than in general, which take a few weeks. This processing requires complex data structures to describe biomolecules. The Python and C languages have been selected for being high-level interpreted languages and performance-efficient compiled languages. This is a nice combination for complex and high-performance-demanding simulations.

Python is selected instead of TCL and Perl because it has a number of desired features, such as integration with compiled languages, library support, object-oriented programming style, and readability.

The user of the MMTK library accesses it as pure Python library, as the C language code used to develop MMTK is completely written in the form of a Python extension module. This code is written only for the time- and performance-critical aspects of the library. For example, interaction energy evaluation is a time-critical function, and energy minimization and molecular dynamics are iterative processes that take much more time to complete; they require high-performance processing. These functions are implemented in the C language to avoid overhead of the Python language. On the other hand, MMTK extensively uses numerical Python, LAPACK, and the NetCDF Python libraries. MMTK also supports shared memory parallel processing (using multithreading) and distributed memory parallel processing (using MPI).

MMTK is designed with the purpose of making it highly extensible. There is no need to modify the MMTK code to add algorithms, energy terms, and specialized data types. For visualization-related activities, MMTK depends on external tools; VMD and PyMOL are especially integrated with MMTK. Generally, MMTK users use Python scripts to access the library. However, several programs use MMTK with a graphical user interface, for example, the DomainFinder and nMOLDYN programs.

MMTK is composed of three major categories of classes. The largest category is the set of classes used to represent atoms and molecules and the classes that manage the database of molecules and fragments. There is a separate subclass of the generic molecule class for representing biomolecules such as DNA, proteins, and RNA. The second (and important) part of MMTK implements various schemas to calculate interaction energies. The third part of MMTK handles input- and output-related functionality. This code is designed to perform read and write functionality for some popular file formats and a custom MMTK format based on the NetCDF format. MMTK's files are portable across various platforms and they are binary in type. As these files are binary files, they are smaller in size and allow efficient access.

Standard Python packages

Besides these, there are specific tools, APIs, and applications available; you can visit the Python Package Index, available online at `http://pypi.python.org`. It contains thousands of modules (mostly developed in Python) for specific applications. These applications again cover a number of scientific, commercial, and computational domains. There are modules for bioinformatics, healthcare, geospatial analysis, instrumentation, engineering, mathematics, and other branches. The web portal maintains a category-wise list of packages. The following list contains some selected packages from science and engineering domains:

- `fluiddyn`: A framework for studying fluid dynamics
- `DeCiDa`: Device and circuit data analysis
- `python-vxi11`: A Python VXI-11 driver for controlling instruments over the Ethernet
- `pygr`: This is a Python graph-database toolkit oriented primarily toward bioinformatics applications
- `Brainiac`: These are various components for use in an artificial intelligence system, and they are entirely usable on their own
- `pyephem`: This computes the positions of planets and stars
- `PyMca`: This is an X-ray fluorescence analysis toolkit and application
- `openallure`: A voice-and-vision-enabled dialog system
- `BOTEC`: A simple astrophysics and orbital mechanics simulator
- `pyDGS`: Wavelet-based digital grain size analysis
- `MetagenomeDB`: A database of metagenomic sequences and annotations
- `biofrills`: These are bioinformatics utilities for molecular sequence analysis
- `python-bioformat`: Read-and-write file formats for life sciences

- `psychopy_ext`: A framework for a rapid reproducible design, analysis, and plotting of experiments in neuroscience and psychology
- `Helmholtz`: A framework for creating neuroscience databases
- `pysesa`: PySESA is an open source project dedicated to providing a generic Python framework for spatially explicit statistical analyses of point clouds and other geospatial data in the spatial and frequency domains for use in geosciences
- `nitime`: Time series analysis for neuroscience data
- `SpacePy`: Tools for space science applications
- `Moss`: Statistical utilities for neuroimaging and cognitive science
- `cclib`: Parsers and algorithms for computational chemistry
- `PyQuante`: Quantum chemistry in Python
- `phoebe`: Physics of stars and stellar and planetary systems
- `mcview`: A 3D/graph event viewer for High-energy Physics event simulations
- `yt`: An analysis and visualization toolkit for astrophysical simulations
- `gwpy`: A package for enable gravitational wave astrophysics in Python

Summary

In this chapter we discussed several case studies of scientific computing applications, libraries, and tools developed using the Python language. We discussed the applications of Python in various areas, including the designing of specialized software and hardware (such as OLPC and ExpEYES) and also designing Python-based embedded systems for lighting systems. We covered some uses of Python in computational chemistry and molecular modeling. We also saw the uses of Python in computer-aided modeling for science and other areas.

In the next chapter, we will discuss the best practices to be followed for the development of scientific computing applications and APIs in general, with a special focus on Python.

10
Best Practices for Scientific Computing

This chapter discusses the best practices to be adopted by developers of applications, APIs, and tools for scientific computing. Best practices are well-defined processes/statements established through research and experience. Following these practices leads to achievement of the desired results with less effort and fewer problems.

In particular, the following topics will be covered in this chapter:

- Best practices for Python-based designing
- Best practices for implementation in Python
- Best practices for data management and application deployment
- Best practices for achieving high performance
- Best practices for privacy and security
- Best practices for maintenance and support
- Best practices for Python development

Generally, scientists use computing tools to support their research, and most of them would not have taken formal computer science training. This may lead them to develop inefficient solutions, and the development cycle might be longer and more time-consuming for them. Also, there might be the possibilities that the implemented algorithms are non-optimal, the development time is longer, and the code is not up to the desired standards. Best practices help them cope up with these problems. Following best practices lets scientists practice science as a proper approach to software development and keeps their code free of unwanted bugs/errors.

There are a number of scientific libraries/applications/toolkits completely designed and developed by scientists from non-computer-science backgrounds. Still, these best practices help them achieve better results and improve the efficiency and overall experience of development.

Best practices can be considered a repeatable standard approach of performing the desired tasks in software development.

The best practices for designing

This section covers the best practices to be followed during the design phase of software development:

- **Task division among different teams**: This best practice says that it is better to distribute the activities of the different steps of the development life cycle among different people. This will reduce the burden on one person and achieve better results in less time. It is better to choose one team (maybe one to two members per team) for each of the design, implementation, and testing phases. These teams will work in collaboration with each other, as they have their respective responsibilities to be performed at different times. This is always better than different teams collaborating with each other. This collaboration is depicted in the following figure with common shape in different steps. The designers may support the developer team in programming, and this approach can be considered as pair programming. Similarly, the collaboration of an implementation team and a testing team will result in fixing of future bugs, and ultimately improve the overall performance of the system.

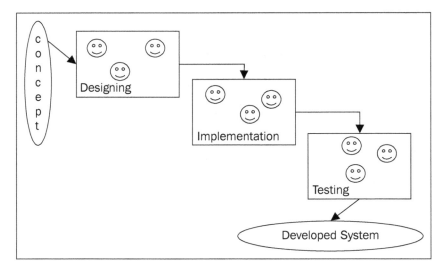

- **Dividing the large task into a set of smaller tasks**: Instead of writing a large program once to perform a large activity, we should prefer dividing the task into smaller subtasks. This is an incremental approach that proceeds toward achieving large tasks by completing smaller subtasks one by one. This will improve the overall implementation experience and the quality of the implemented code. Following this approach will produce better code that requires less development efforts and is easily manageable.

- **The life cycle for each subtask**: Following the development life cycle (that is, designing, developing, and testing) for each small subtask will definitely produce code that is less prone to errors, as each subtask is very well tested. This will also improve the overall quality of the code as the team will be dealing with smaller pieces of code. This approach is presented in the following figure:

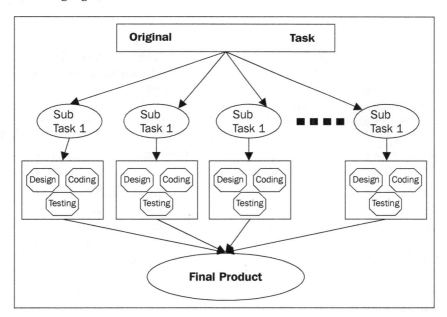

Life cycle of each sub task

Adopting this approach will prevent the developer team from suddenly getting trapped in major errors. This will also save efforts during the final testing of the complete application.

- **Use specialized software for each activity**: It is recommended that teams use specialized software for each activity of the development life cycle. There are various pieces of standard software for most activities. For example, there are pieces of software for designing and modeling tools (for example, Visio). To support development activities, we have integrated development environment software (for example, Eclipse for Java), version control software (for example, CVS or Git), debuggers (for example, GDB), and build tools (for example, ANT). There are also specific pieces of software for testing of applications and performance profiling.

The implementation of best practices

This section covers the best practices to be adopted for the implementation phase of software development:

- **Maximize comments in the code and documentation**: Most scientific applications involve complex algorithms and computations; hence, their implementation is also complicated. It will be better for future enhancements if most complex implementations have descriptive comments to explain what the code is doing. It is necessary to maximize comments and documentation so that the users/developers are well aware of the ideas behind the programs. Specially, proper comments with complex logic will enable the development team working on future enhancement of the application/tool/ API developed. The comments are supposed to explain the logic of the code.

- **Promote reusability**: Instead of reinventing the wheel, before starting the development cycle, search for suitable libraries for the purpose. This will save a lot of effort by developing some already existing libraries. Moreover, the code developed using existing and well-tested libraries will have a much lower possibility of runtime errors, or bugs, as these libraries have supposedly been already tested and used a number of times. Using an existing library will keep the scientist's mind free to immerse in science. This will save a lot of effort; the only effort required are those of learning about the library and using it for the task to be performed.

- **Develop a complete working model first**: A good way of developing an application, tool, or API is to first develop a working model and then plan to optimize the solution developed. This is the approach that should be followed for even simple commercial applications. Optimization can be performed on the working model to improve its performance. However, planning for optimization along with development may divert the attention of the team. Hence, the focus during the development phase should be on achieving the desired functionality, and optimization can be applied on the properly working application, tool, API to improve its performance.

- **Consider the possibility of future errors**: Adopt a proactive approach to cope up with future errors. This approach involves the use of assertions, exception handling, automated tests, and debuggers. Assertions can be used to ensure that preconditions and postconditions of a particular piece of code are intact. An automated test helps the developer ensure that the behavior of the program remains unchanged, even with modifications in the program. Each of the errors detected during testing should be converted into a test case, so that it will be automatically tested in the future. The use of debuggers is always a better choice than inserting the `print` statement to test the validity of the program. Using debuggers will help the developer get an insight into the impact of each statement of the program. Exception handling helps developers handle possible errors proactively. The following code fragment demonstrates the use of assertions in Python:

```
# assertion for precondition testing
def centigradeToFahrenheit (centigrade)
assert type(centigrade) is IntType, "Not an integer"
  assert (centigrade >= 0), "Less then absolute Zero"
  return (9 * centigrade/5 + 32)
print centigradeToFahrenheit(40)
print centigradeToFahrenheit(15)
print centigradeToFahrenheit(-10)
# assertion for both pre and post condition testing
def calculate_percentage (marks1, marks2, marks3)
  assert (marks1 >= 0), "Less then absolute Zero"
  assert (marks2 >= 0), "Less then absolute Zero"
  assert (marks3 >= 0), "Less then absolute Zero"
  result = (marks1 + marks2 + marks3)/100.0
  assert (0.0 <= result <= 100) "Percentage should be
    between 0 and 100"
  return result
```

- **Open source/standard publication of data and code**: Code development and the preparation of data for experimentations should ideally be done with the aim of publishing them as open source/standards so that both the code and the data will be adopted by peer scientists working in the same field. This will increase awareness about the application, tool, or API, and ultimately result in a large user base.

Publishing the data and code increase the user base and these users will support the efforts in testing and future enhancement. The data will also be improved and will keep getting updated with new user requirements. Generally, open source software is updated in collaboration with a number of developers and scientists from various geographical locations. To support a large number of distributed developers, a recent trend in version control software is distributed version control. Distributed version control is a web-based system that is highly scalable for supporting a large number of developers. Traditional version control software was not designed to support a very large number of users working on the code in the software repository.

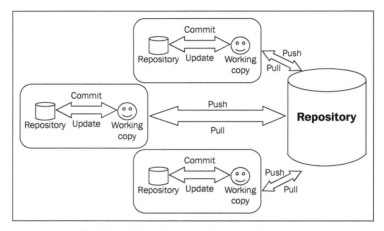

Working of distributed version control software

The best practices for data management and application deployment

This section covers the best practices for data management and deployment of applications:

- **Data replication**: This practice especially focuses on mission-critical applications, where data loss is intolerable, which may be due to the high cost of the experiment, or where the experiment's failure can result in a loss of life. For such mission-critical applications, data replication should be properly planned such that a failure of some components of the system will not affect the overall functionality of the system. The replicated data must be placed at different locations so that a natural disaster at one location will not affect the ultimate processing.

The following figure depicts the concept of data replication. Each piece of data is replicated three times at different locations across the globe. Even if there is a failure of one or two systems that have a particular piece, the processing doesn't stop as there is another copy.

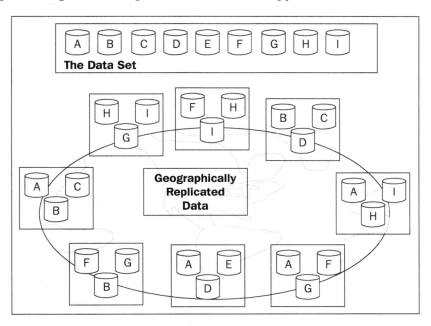

- **Testing on real and synthetic data**: Testing of the application should be done on real as well as synthetic data. If real test data is not available, then perform tests on synthetic data. To prepare synthetic data, use the appropriate statistical-distribution-based random number generation techniques, as discussed in *Chapter 3, Efficiently Fabricating and Managing Scientific Data*. Generally, there are open datasets for most common scientific applications, as pointed in *Chapter 3, Efficiently Fabricating and Managing Scientific Data*. If found suitable, these datasets may be used for experimentation and testing of the application.

The best practices to achieving high performance

This section is specially focused on applications that require high performance. The best practices for achieving high performance are presented in this section:

- **Consider the future scalability requirements**: It is better to proactively consider the future scalability requirements of the system. The data size for the system may vary from very small data to a huge dataset up to a "Petascale" or "Exascale"; during designing, this aspect must be considered during design. Depending on the requirements, the hardware infrastructure, the software development framework, and the database may be planned. This design process should consider the possibility that the system may require the process huge datasets in future.

- **Hardware and software selection**: Invest a sufficient amount of time to selecting the most suitable technology for the application, tool, or API. This process will require an investment of initial efforts to select a suitable environment that will result in a satisfactory implementation of the desired functionality. This selection of technology includes choosing a suitable programming language and development framework, the appropriate database/data store, the required hardware, a suitable deployment environment, and others.

- **API selection**: If there are some existing APIs for achieving the desired functionality, then the selection of the most suitable APIs for achieving the desired result is crucial for a successful and efficient implementation. Before you finalize the API to be used, it should be properly analyzed for the functional and performance requirements. The ultimate performance of the final product is directly dependent on the API used to build the system.

- **Use the appropriate performance benchmarks**: For performance-critical applications, use the appropriate performance benchmarks. There are a number of benchmarks available for assessing the performance of different types of applications, tools, and APIs. For example, the DEISA benchmark suite is a specially designed high-performance scientific computing application. Generally, a benchmark is composed of a set of custom or real programs from the application domain. These programs will be executed several times to assess the performance of the system under study.

The best practices for data privacy and security

Data privacy and security are the most important areas to focus on for applications to be successfully adopted and widely used. This section covers the best practices for proper privacy and security of the application and data:

- **Data privacy**: There are certain applications that require collection of data, and for some specific applications, the developers must take utmost care of the privacy of user data. This data privacy is highly essential, as the data may be financial or medical data, and losing it in some way may result in a big loss for the stakeholders. This concern must be considered carefully during all phases of the system's development life cycle.

- **Security considerations for web applications/services**: If the application is designed as a web application/service, then special care must be taken when it comes to security, as web-based systems are the main focus of security attacks. Several well-defined strategies are available for both securing and attacking web-based systems. Preventive measures should be considered from the very first step of the system development life cycle for the application. Proper authorization and authentication mechanisms should be adopted to achieve both privacy and security for the application.

Testing and maintenance best practices

Proper testing and maintenance are highly essential for proper software development. This section emphasizes the best practices to be followed during the testing and maintenance activities:

- **Unit testing first**: It will be better to perform unit-wise testing first. After successful unit testing, the system is ready for integration testing. Finally, after successful integration testing, validation testing should be performed. Unit testing ensures that the different modules of the system are working perfectly and helps in early detection of errors. This will not only fix the bugs in the module, but also support in finding the missing parts of the implementation of the original idea. As unit testing is performed for a specific module at a time and the focus is very small, it may identify the parts of the specification that were missed during the implementation phase.

- **Different testing teams**: Testing is a crucial activity for the success of the final product. It will be better to have different teams for different functions of the testing phase. These teams may work in collaboration for better results. This will help identify bugs and issues in the implementation and give an overall better quality to the final product.

- **Working groups for support**: To provide proper support and maintenance activities for a large system, it is better to create multiple working groups for each of the substantial subparts of the system. It is preferable that at least one of the coders of that specific subtask also be a member of its support group. In this way, this coder (and member) may easily identify the problem and fix it. Each working group is responsible for a specific subpart. In this way, each member of the group will have a thorough idea of that subtask. These members will easily manage the support and maintenance activity.

- **Multiple working groups**: For large applications, create multiple working groups to divide the workload. Each work group is responsible for providing support, maintenance, and enhancement activities. A dedicated working group for a specific module will help improve the overall quality of the system and give better support. As the team comes across problems related to a specific module in a short time span, they will understand the problems associated with the system and finally recommend the desired update to the system.

- **Mailing lists for user help and support**: Create a user mailing / feedback list for each working group. Users will raise the problems to this mailing list, and members of the support team will respond with the solution on the mailing lists. This list will serve as a communication bridge between users and developers.

General Python best practices

This section discusses some general best practices that should be followed by Python programmers:

- **The PEP 0008 style guide for Python code**: The first best practice in this category is to clearly understand and follow PEP 0008. Refer to `https://www.python.org/dev/peps/pep-0008/`.

- **Naming convention**: It is recommended to all coders that they follow a consistent and meaningful naming convention. This recommendation is helpful, not only to the original developers, but also to future developers who may work to enhance the system. Uniform and meaningful names improve the readability of code. The naming convention should follow a uniform naming scheme and adopt the recommended scheme for the specific language under use, for example, the use of underscores or camel case to join multiple words in a variable or function name. The following table represents the recommended names and the not-recommended names:

Not recommended	Recommended
Variables:	Variables:
var1, var2, mycalculation,	area, incomeTax, productCost,
temp_val,	counter,
f1, num35	lambda, sigma,
	sum_of_product
Functions:	Functions:
func1(), function2(), calculation_func(),	calculateArea(), product_of_sum()
perform_func()	sinx()

- **Uniform coding style**: Generally, it is recommended that you follow a standard and uniform coding style throughout the system. The use of assertions, indentation, comments, and other things must be uniform in the code. Adopt or develop a standard style for comments and follow it in the coding throughout the system. Similarly, formatting should also be the same in the entire code for the system. It should consider spacing and indentations in the code.

The following example depicts the poor and recommended formatting styles:

Nonuniform spacing and indentation	Uniform spacing and indentation
x=(b*d- 4*a*c)/2*a	x = (b * d – 4 * a * c) / 2 * a
y = 2 * x * x + 4 * x + 5	y = 2 * x * x + 4 * x + 5
def sample_function()	def sample_function()
print "in function"	print "in function"
print " last line"	print " last line"
def second_sample()	def second_sample()
print "in function"	print "in function"
print "last line"	print "last line"

Summary

In this chapter, we have discussed best practices to be followed by the teams working on scientific computing. The chapter started with a discussion on best practices for designing. After that, the best practices for coding were discussed. Later, the best practices for data management and application deployment were covered.

Next, the best practices for high-performance computing were discussed, and then the best practices for security and data privacy were presented. Then, we saw the best practices for maintenance and support. Finally, the best practices of general Python-based development were discussed.

Index

C

Calculus 138-140
chart plotting library, Python 37
classes, associated with arrays
  about 95
  masked array 96
  matrix sub class 96
  structured/recor array 97
clustering analysis 71
comma-separated values (CSV) 186
Common Data Format (CDF) 48
compact disks (CDs) 62
complex numbers 135
computational chemistry applications 245
Computer Algebra System (CAS) 76
computer arithmetic 11
conditioning 10
cryptography module 150, 151
CSV files
  working on 186-189
curve fitting 126, 127

D

data
  about 41
  scientific example 42
database 45
DataFrame, pandas library 171
data management and application
    deployment
  best practices 258, 259
data management, operations
  data architecture, analysis, and design 45
  database administration 45
  data enrichment 46
  data governance 45
  data integration 46
  data integrity 46
  data quality management 46
  data security management 46
  data warehouse management 46
  metadata management 46
data structures, pandas
  DataFrame 84
  Panel 84
  Series 84

data warehouse management 46
DICOM 49
differential equations
  about 25
  boundary value problem 26
  initial value problem 26
dsolve method 138

E

energy-efficient lights
  with embedded system 248, 249
equation solving 133, 134
error analysis 10
European data format 49
exemplary programs
  about 81
  simple integrations 83
  simplification, of expression 82
  simplification, of formula 82
  symbol manipulation 82
Experiments for Young Engineers and
    Scientists (ExpEYES) 236-239
exponentials
  functions 134
expressions 132
extraction, translation, and loading
    (ETL) 46
extrapolation
  about 23
  methods 24

F

File I/O (scipy.io) 128, 129
files
  about 43
  structured files 44
  unstructured files 44
financial charts 40
FITS astronomical data 49
five-digit random numbers
  generating, logic 62
Flexible Image Transport System (FITS) 48
floating-point numbers 11
ForecastWatch
  about 239
  components 239, 240

P

pandas
about 83
data structures 84
features 84
used, for data analysis 83
used, for manipulation 83
pandas library
about 170
common functionalities, among data
structures 174-179
DataFrame 171
missing data, handling 184
Panel 172
Series 170
time series and date functions 181-183
pandas plotting 190, 191
PanMachine 247
partial differential equations (PDE)
about 26
methods, for solving 26
Pauli algebra 142
physics module
about 142
high-energy physics 145
Hydrogen wave functions 142
matrices 142
mechanics 146, 147
Pauli algebra 142
second quantization 143
Physics with Homemade Equipment
and Innovative Experiments
(PHOENIX) 236
pie charts 39
platforms, Sugar
Live CD and Live USB stick 235
OS image 235
package for Linux distributions 235
XO laptop 235
plotting module
functions, using 79
polar plots 39
polynomial manipulation 158, 159
polynomials 135
PowerPanel 247

pretty printing
about 148, 149
LaTeX Printer 149
privacy and security
best practices 261
Product Information Model (PIM) 249
product of sum (POS) 154
pseudo-random number generators 27, 28
pyACDT API 240
PyDrone 245
PySpark 232
Python
background 12
best practices 262, 264
drawbacks 17
for developing Blind Audio Tactile
Mapping System (BATS) 246
guiding principles 13, 15
scientific computing applications 234
scientific computing libraries 249
URL, for guiding principles 13
Python aircraft conceptual design toolbox
(pyACDT) 241
Python Enhancement Proposals (PEP) 15
Python, for scientific computing
about 15
available libraries 17
compact and readable code 15
data structures 16
free and open source 16
graphical user interface packages 16
hierarchical module system 16
holistic language design 15
language interoperability 16
portable and extensible 16
testing framework 17
Python Package Index
about 251
reference link 251
Python's built-in functions, using for
random number generation
about 55
bookkeeping functions 56
functions, for integer random number
generation 56

V

VAMPzero tool 240
vectors 141

W

WatchMan 244
Web ARChive (WARC) format 230
wxPython 247

Thank you for buying
Mastering Python Scientific Computing

About Packt Publishing

Packt, pronounced 'packed', published its first book, *Mastering phpMyAdmin for Effective MySQL Management*, in April 2004, and subsequently continued to specialize in publishing highly focused books on specific technologies and solutions.

Our books and publications share the experiences of your fellow IT professionals in adapting and customizing today's systems, applications, and frameworks. Our solution-based books give you the knowledge and power to customize the software and technologies you're using to get the job done. Packt books are more specific and less general than the IT books you have seen in the past. Our unique business model allows us to bring you more focused information, giving you more of what you need to know, and less of what you don't.

Packt is a modern yet unique publishing company that focuses on producing quality, cutting-edge books for communities of developers, administrators, and newbies alike. For more information, please visit our website at www.packtpub.com.

About Packt Open Source

In 2010, Packt launched two new brands, Packt Open Source and Packt Enterprise, in order to continue its focus on specialization. This book is part of the Packt Open Source brand, home to books published on software built around open source licenses, and offering information to anybody from advanced developers to budding web designers. The Open Source brand also runs Packt's Open Source Royalty Scheme, by which Packt gives a royalty to each open source project about whose software a book is sold.

Writing for Packt

We welcome all inquiries from people who are interested in authoring. Book proposals should be sent to author@packtpub.com. If your book idea is still at an early stage and you would like to discuss it first before writing a formal book proposal, then please contact us; one of our commissioning editors will get in touch with you.

We're not just looking for published authors; if you have strong technical skills but no writing experience, our experienced editors can help you develop a writing career, or simply get some additional reward for your expertise.

Learning SciPy for Numerical and Scientific Computing
Second Edition

ISBN: 978-1-78398-770-2   Paperback: 188 pages

Quick solutions to complex numerical problems in physics, applied mathematics, and science with SciPy

1. Use different modules and routines from the SciPy library quickly and efficiently.

2. Create vectors and matrices and learn how to perform standard mathematical operations between them or on the respective array in a functional form.

3. A step-by-step tutorial that will help users solve research-based problems from various areas of science using Scipy.

IPython Interactive Computing and Visualization Cookbook

ISBN: 978-1-78328-481-8   Paperback: 512 pages

Over 100 hands-on recipes to sharpen your skills in high-performance numerical computing and data science with Python

1. Leverage the new features of the IPython Notebook for interactive web-based big data analysis and visualization.

2. Become an expert in high-performance computing and visualization for data analysis and scientific modeling.

3. A comprehensive coverage of scientific computing through many hands-on, example-driven recipes with detailed, step-by-step explanations.

Please check **www.PacktPub.com** for information on our titles

Mastering Object-oriented Python

ISBN: 978-1-78328-097-1   Paperback: 634 pages

Grasp the intricacies of object-oriented programming in Python in order to efficiently build powerful real-world applications

1. Create applications with flexible logging, powerful configuration and command-line options, automated unit tests, and good documentation.

2. Use the Python special methods to integrate seamlessly with built-in features and the standard library.

3. Design classes to support object persistence in JSON, YAML, Pickle, CSV, XML, Shelve, and SQL.

Python High Performance Programming

ISBN: 978-1-78328-845-8   Paperback: 108 pages

Boost! the performance of your Python programs using advanced techniques

1. Identify the bottlenecks in your applications and solve them using the best profiling techniques.

2. Write efficient numerical code in NumPy and Cython.

3. Adapt your programs to run on multiple processors with parallel programming.

Please check **www.PacktPub.com** for information on our titles

www.ingramcontent.com/pod-product-compliance
Lightning Source LLC
Chambersburg PA
CBHW060516060326
40690CB00017B/3302